Assessment of TOC Analytical Accuracy

The mission of the Awwa Research Foundation (AwwaRF) is to advance the science of water to improve the quality of life. Funded primarily through annual subscription payments from over 1,000 utilities, consulting firms, and manufacturers in North America and abroad, AwwaRF sponsors research on all aspects of drinking water, including supply and resources, treatment, monitoring and analysis, distribution, management, and health effects.

From its headquarters in Denver, Colorado, the AwwaRF staff directs and supports the efforts of over 500 volunteers, who are the heart of the research program. These volunteers, serving on various boards and committees, use their expertise to select and monitor research studies to benefit the entire drinking water community.

Research findings are disseminated through a number of technology transfer activities, including research reports, conferences, videotape summaries, and periodicals.

Assessment of TOC Analytical Accuracy

Prepared by:
Louis A. Kaplan
Stroud Water Research Center
Avondale, PA 19311

George R. Aiken
Water Resources Division
United States Geological Survey
Boulder, CO 80303

Preston Luitweiler
Philadelphia Suburban Water Company
Bryn Mawr, PA 19010

Sponsored by:
Awwa Research Foundation
6666 W. Quincy Avenue
Denver, CO 80235

Published by the

Awwa Research Foundation
Advancing the Science of Water®

IWA Publishing

DISCLAIMER

Published by IWA Publishing, Alliance House, 12 Caxton Street, London SW1H 0QS, UK

Telephone: +44 (0) 20 7654 5500; Fax: +44 (0) 20 7654 5555; Email: publications@iwap.co.uk;
Web: **www.iwapublishing.com**

AwwaRF report number 90929F. Originally published by AwwaRF for its subscribers in 2003.
IWA Publishing version published 2004.

ISBN 1 84339 845 1

CONTENTS

TABLES

FIGURES

FOREWORD

The Awwa Research Foundation is a nonprofit corporation that is dedicated to the implementation of a research effort to help utilities respond to regulatory requirements and traditional high-priority concerns of the industry. The research agenda is developed through a process of consultation with subscribers and drinking water professionals. Under the umbrella of a Strategic Research Plan, the Research Advisory Council prioritizes the suggested projects based upon current and future needs, applicability, and past work; the recommendations are forwarded to the Board of Trustees for final selection. The foundation also sponsors research projects through the unsolicited proposal process; the Collaborative Research, Research Applications, and Tailored Collaboration programs; and various joint research efforts with organizations such as the U.S. Environmental Protection Agency, the U.S. Bureau of Reclamation, and the Association of California Water Agencies.

This publication is a result of one of these sponsored studies, and it is hoped that its findings will be applied in communities throughout the world. The following report serves not only as a means of communicating the results of the water industry's centralized research program but also as a tool to enlist the further support of the nonmember utilities and individuals.

Projects are managed closely from their inception to the final report by the foundation's staff and large cadre of volunteers who willingly contribute their time and expertise. The foundation serves a planning and management function and awards contracts to other institutions such as water utilities, universities, and engineering firms. The funding for this research effort comes primarily from the Subscription Program, through which water utilities subscribe to the research program and make an annual payment proportionate to the volume of water they deliver and consultants and manufacturers subscribe based on their annual billings. The program offers a cost-effective and fair method for funding research in the public interest.

A broad spectrum of water supply issues is addressed by the foundation's research agenda: resources, treatment and operations, distribution and storage, water quality and analysis, toxicology, economics, and management. The ultimate purpose of the coordinated effort is to assist water suppliers to provide the highest possible quality of water economically and reliably. The true benefits are realized when the results are implemented at the utility level. The foundation's trustees are pleased to offer this publication as a contribution toward that end. Nearly all aspects of research on natural organic matter (NOM) involve the analysis of total organic carbon (TOC) as a surrogate for NOM. With the recent promulgation of the disinfectant/disinfection by-product (D/DBP) rule, TOC has taken on new significance for the drinking water industry, as TOC measurements are a basis for compliance with U.S. EPA regulations. Utilities are required to measure TOC concentrations in treatment plant influents, waters that contain a variety of dissolved molecules and organic particles. Uncertainty about the analytical accuracy of TOC measurements is the result of inadequate proficiency evaluation standards. This report evaluates several potential TOC standards that can allow the utility analyst to assess performance with both difficult to oxidize solutions and organic particles, all within a complex inorganic matrix.

Edmund G. Archuleta
Chair, Board of Trustees
Awwa Research Foundation

James F. Manwaring, P.E.
Executive Director
Awwa Research Foundation

ACKNOWLEDGMENTS

The authors of this report are indebted to the water utilities, TOC instrument manufacturers, academic laboratories, and government laboratories for their cooperation and participation in this project. Sherman L. Roberts, Amy Law, and Michael Gentile provided technical assistance. The authors thank the PAC members Gordon Brand, Ken Carlson, Ph.D., and Hisao-Chiu Wang for constructive comments that improved this project.

EXECUTIVE SUMMARY

BACKGROUND

Natural organic matter (NOM) is a component of all water supplies and its presence has implications for drinking water utilities beginning with watershed management and carrying through water treatment to water distribution. Two of the more noteworthy aspects of NOM for water utilities are the production of disinfectant-by products (DBP) resulting from the reaction of NOM with disinfectants and the growth of bacteria on biodegradable organic matter (BOM). Nearly all aspects of research on natural organic matter involve the analysis of total organic carbon (TOC) as a surrogate for NOM. With the recent promulgation of the disinfectant/disinfection by-product (D/DBP) Rule, TOC has taken on new significance for the drinking water industry. TOC measurements may be the basis for instituting potentially costly changes in treatment operation and facilities to insure compliance with United States Environmental Protection Agency (U.S. EPA) regulations regarding TOC removal.

Utilities are required to measure TOC concentrations in treatment plant influents, waters that contain a variety of dissolved molecules and organic particles. Uncertainties about the analytical accuracy of TOC measurements are the result of inadequate proficiency evaluation standards.

The major activity of this project was to address three goals: (1) to develop TOC standards that can be used with all TOC methods to evaluate performance, (2) to determine the ability of different TOC methods to accurately measure forms of TOC in complex environmental matricies, and (3) to test the ruggedness of the proficiency standards in a round-robin survey involving a large number of laboratories. Additionally, we addressed the question of whether TOC or dissolved organic carbon (DOC) is the more appropriate measure of reactive NOM when the ultimate issue is regulation of DBP formation. This question was addressed in our fourth research goal: (4) to determine the contribution of natural organic particles, suspended organic carbon (SOC), in the formation of DBP's. This report presents our evaluation of several potential TOC standards that can allow the utility analyst to assess performance with both difficult to oxidize solutions and organic particles, all within a complex inorganic matrix.

APPROACH

The research project was coordinated through the Stroud Water Research Center. All standards were prepared at the Stroud Center and distributed to each of the project participants. Close collaboration with one government laboratory from the United States Geological Survey (USGS), Water Resources Division laboratory in Boulder, Colorado, under the direction of Dr. George Aiken, helped in the selection of materials to be screened for standards, contributed to the analysis of those materials, and helped identify what standards would be distributed to the other participants during the initial screening of candidate compounds.

Thirty-two participating laboratories were sent standards three different times during the project. Each shipment consisted of a series of 8 standards, each supplied in triplicate 40-mL vials. Standards prepared in a deionized water matrix were the focus during the first exchange, followed by a focus on inorganic carbon and mineral salt matrix modifications for subsequent sets. In the second and third exchanges, the identity of the previous standards and the results from all laboratories were shared with the analysts. The standards included potassium hydrogen phthalate (KHP), caffeine, dodecylbenzene sulfonic acid, humic acid, fulvic acid, leaf powder, polystyrene microbeads, and freeze dried bacteria. Each participant received a data sheet for reporting, was given a code, and was asked to analyze and return the data within a 4- to 6-week period. Cover letters accompanied each shipment and provided some information to help the analyst, such as identifying the standards containing suspensions. One shipment included an analyst questionnaire designed to reveal procedural aspects of each laboratory.

A hierarchical approach was followed in the design of the standards testing and data analyses. Specifically, we considered individual machines, analysts, or operators as the lowest test level, and as nested within instrument models. Analysts and instrument model, in turn, were nested within one of three oxidation methods, Uv-persulfate, Pt-persulfate, and high-temperature combustion. To the extent that from within the participant pool, more than one operator used the same instrument model, or that more than one instrument model fell under an oxidation method, we were able to ascribe the error associated with the analyses of standards to method, instrument, or machine (i.e. operator or analyst) within this hierarchy.

The influence of particles on disinfection byproduct formation potential (DBPFP) was assessed in cooperation with Philadelphia Suburban Water Company (PSWC). A protocol developed by PSWC for a DBPFP test was applied to samples from two different water sources (Crum Creek and Neshaminy Creek) under two different treatments, unfiltered and filtered. Crum Creek has water drawn from a reservoir within a watershed dominated by agricultural uses. The Neshaminy Creek water comes from a watershed dominated by residential and industrial land uses, and receives wastewater treatment discharges. PSWC personnel collected the water samples and Stroud Center personnel filtered the water. Samples were taken for the DBPFP test, the measurement of TOC, ash free dry mass (AFDM), and chlorophyll a. The test protocol developed by PSWC resulted in a characterization of the chlorine demand of the water samples and a determination of DBPFP for the waters at two applied chlorine doses (one dose close to the operating applied dose, and one close to a target chlorine dose that is 0.5 to 1.0 mg/L above the breakpoint) at uniform incubation conditions of time, temperature, and pH. Each sample was analyzed at PSWC for the following DPB's: (1) haloacid (monochloroacetic acid, monobromoacetic acid, dichloroacetic acid, dibromoacetic acid, trichloroacetic acid, and bromochloroacetic acid); (2) pentane extractable trihalomethanes (THM) and haloacetonitriles (HAN) (chloroform, bromoform, bromodichloromethane, dibromochloromethane, haloacetonitrile, trichloroacetionitrile, dichloroacetonitrile, dibromoacetonitrile, bromochloroacetonitrile); and (3) other compounds (dichloroacetone, trichloroacetone, chloropicrin, and chloral hydrate).

RESULTS

In general, all laboratories performed well in the analysis of KHP, most performed well in the analyses of difficult to oxidize solutions of organic molecules, some laboratories had problems separating inorganic carbon from organic carbon, and nearly all laboratories had difficulty with standards containing particle suspensions. A useful context for the TOC evaluation is the U.S. EPA certification range of ± 20% accuracy. The average recovery, defined as the ratio of the measured concentration to the theoretical concentration, expressed as a percent, for the 39 machines used to measure KHP was 103 ± 7% ($\bar{x}$ ± SD), with a range of 86 to 125%. All but one of the machines would have qualified under the U.S. EPA guidelines of ± 20% accuracy, with the next lowest recovery equal to 86%. Thirty-three of the machines or analyzers (85% of the total) were operated within 10% of the theoretical value, with 23 of them within 5% or better. When separated by the 12 instrument models, the range of recoveries was 96 to 113%. There was little influence of mineral salts on the analysis of KHP. In contrast, several analysts had difficulty in efficiently separating an inorganic C (IC) amendment from the KHP standard when a 30 mg IC/L spike as added to the standard.

Dodecylbenzene sulfonic acid and caffeine were particularly difficult molecules for numerous analysts, and might be good choices as proficiency evaluation standards. For dodecylbenzene, 15 of the 39 analysts achieved less than 80% accuracy. The average recovery was 83 ± 20%, and the range was from 43 to 110%. Nine of 39 analysts achieved less than 80% accuracy with caffeine. The average recovery for all analysts was 89 ± 14%, and the range was 63 to 114%. Naturally occurring compounds, such as humic and fulvic acid, were not as problematical for the analysts, with an average recovery of 92 ± 12% for humic acid, and 92 ± 9% for the humic/fulvic acid mixture. Two of the analysts achieved less than 80% accuracy with the humic/fulvic acid mixture (range of recoveries 65 to 109%), and 5 of the analysts achieved less than 80% accuracy with the humic acid alone (range of recoveries 71 to 117%).

Particles were difficult for nearly all analysts, with the average recoveries and ranges equal to 70 ± 9% (range 54 to 100%) for the leaf powder, 52 ± 31% (range 10 to 132%) for P-17, and 42 ± 34% (range 2 to 116%) for the polystyrene polybeads. The high temperature instruments, on average, performed better with particles than instruments of all oxidative classes, but all instrument classes had a broad range of performances. The range of recoveries for the high temperature instruments was 58 to 100% with the leaf powder, 15 to 133% with P-17, and 67 to 116% with the polystyrene polybeads. Our data did reveal problems in the delivery of suspended particles to the TOC oxidation zone for several analyzers. This suggests that improving the oxidation of SOCs will not by itself solve the problem of particle analysis.

Averaged over all instruments and analysts, standards that were solutions were analyzed with 92 ± 9% recovery, while standards that were particles were analyzed with 55 ± 21% recovery. The range for the individual analysts with standards that were a solution was 71 to 109%, and 2 of 39 analysts did not achieve 80% accuracy. The range of recoveries with particles was 23 to 121%, only 3 analysts achieved greater than 80% accuracy, and 16 of 39 analysts had less than 50% accuracy. When averaged over the individual instruments, all of the instrument models achieved at least 80% recovery with solutions, while only 2 instrument models achieved better than 80% recovery with particles. Apparently there is a large amount of operator variability within the water supply industry, as shown by the wide range of recoveries with the same instrument model.

Our data suggest that particles can compete with dissolved precursor material in reactions with applied disinfectants and contribute DBP precursors. The role of particles in DBPFP, however, seems to be related to overall chlorine demand. DBP formation was generally reduced when the DBPFP tests were chlorine limited and increased when the tests were precursor limited (i.e., chlorine demand was substantially met). The most dramatic increase in DBP formation due to particles was observed in a stream sample where THM's increased 2-fold and chloral hydrate concentration increased 3.6-fold compared to filtered samples. Additional complexity influencing the role of particles includes the particle sources, season, and concentration, especially during storms that generate additional runoff that is typically high in both DOC and SOC.

CONCLUSIONS

The data obtained with the proficiency evaluation standards confirmed our concerns about certification based upon easily oxidized solutions such as KHP. Our data also suggest that more research is needed to ascertain the role of particles in DBP formation within the context of the entire treatment process. We provide the following recommendations:

- there is a need for appropriate proficiency evaluation standards that can allow the analyst to easily assess their own performance with TOC analyses
- routine TOC instrument evaluation should include inorganic carbon removal, oxidation efficiency of difficult to oxidize solutions, and recovery efficiency of particles
- given the difficulty of handling particles in TOC analysis, the industry should consider the importance of organic particle data and assess alternative methodologies for particle evaluation
- additional research is needed to further explore the role of particles, and address at least two critical questions: (1) do particle-generated DBP's remain particle associated during treatment, and (2) what attributes of particles enhance their role as DBP precursors.

CHAPTER 1
INTRODUCTION

BACKGROUND - STATEMENT OF PROBLEM AND SIGNIFICANCE OF THE RESEARCH

Natural organic matter (NOM) is a component of all water supplies and its presence has implications for drinking water utilities beginning with watershed management and carrying through water treatment to water distribution. Two of the more noteworthy aspects of NOM for water utilities are the production of disinfectant-by products (DBP) resulting from the reaction of NOM with disinfectants and the growth of bacteria on biodegradable organic matter (BOM). Nearly all aspects of research on NOM, including many of the studies funded by the Research Foundation, involve the analysis of total organic carbon (TOC) as a surrogate for NOM. With the recent promulgation of the Information Collection Rule (ICR) and the disinfectant/disinfection by-product (D/DBP) Rule, TOC has taken on new significance for the drinking water industry. TOC measurements may be the basis for instituting potentially costly changes in treatment operation and facilities to insure compliance with United States Environmental Protection Agency (U.S. EPA) regulations regarding TOC removal.

TOC in natural waters is a mixture of dissolved organic carbon (DOC) molecules plus particles and colloids, or suspended organic carbon (SOC). Natural waters also contain inorganic carbon (IC) typically dominated by the bicarbonate ion, and the IC concentration is usually several-fold higher than the TOC concentration. Quantitative measurements of organically bound carbon involve the conversion of organic carbon to CO_2 and the measurement of CO_2 concentration by a nondispersive infrared analyzer, coulometric titrator, or conductivity cell. The analysis can be performed by measuring total carbon (TC), which is the sum of IC and TOC, and subtracting IC from TC to estimate TOC. This reduces instrument precision and sensitivity because one large value is subtracted from another. Alternatively, TOC analysis can begin with the removal of inorganic carbon from the sample, followed by the oxidation of the sample, and ending with detection of the oxidized carbon.

The methods used to oxidize organic carbon include catalysis, heat, chemical oxidants, ultraviolet irradiation, or a combination of these methods. The problems and precautions for each step in the analysis of organic carbon have been recently addressed (Aiken 1992; Kaplan 1992) and the equivalency of persulfate oxidation and combustion oxidation methods for freshwaters has been demonstrated with groundwaters, stream water, and river water (Kaplan 1992; Benner and Hedges 1993). There should be little doubt that many of the present generation of TOC analyzers, when operated properly, are capable of providing accurate and precise values for DOC. A recent paper by Koprivnjak et al. (1995) that reported 20 to 30% lower values for DOC in freshwaters analyzed by the persulfate method, is very misleading. All of the instruments used were several generations old, and no variance terms were reported for the measurements.

There is, however, considerable and legitimate uncertainty about the ability of any of the TOC instruments to accurately measure the suspended organic carbon (SOC) within TOC. Three significant problems contribute to that uncertainty, including: (1) taking a representative sample of the suspension, (2) delivering that suspension through needles, tubing, or valves to an oxidation zone, and (3) oxidizing the suspension completely. As an indication of the importance to TOC accuracy of accurately measuring SOC, consider that 33% of the TOC in 25 forested streams and rivers and 16% of the TOC in non-forested streams and rivers is present as SOC

(Golladay 1997). Additionally, for some of the largest rivers in North America, the contribution of SOC to TOC is even greater, including the Brazos (52%), Mississippi (53%), Ohio (37%), and Missouri (81%) (Malcolm and Durum 1976).

Against this backdrop of analytical uncertainty, the U.S. EPA has qualified laboratories to measure TOC based on performance with a calibration standard dissolved in deionized water. In contrast, measurements for the ICR and D/DBP Rule are being made on unfiltered samples with complex matrices including inorganic carbon, other ions, and turbidity. There is no standardized means of assessing interferences or the efficiency of either DOC or SOC oxidation. The net result is that TOC concentrations reported for the ICR at best may represent a combination of DOC and some unknown fraction of SOC.

Clearly, there is a need for an assessment of TOC analytical accuracy, and for the development of relevant TOC standards. Calibration standards are important, but calibration alone is not adequate to permit drinking water utilities to assess their TOC analytical performance. In the context of TOC measurements, performance includes distinguishing IC from TOC, and efficiently oxidizing recalcitrant DOC and SOC in matrices representative of natural waters.

The analytical difficulties associated with accurately measuring SOC are such that some questioning of the scientific rationale for analyzing TOC rather than DOC is appropriate. With regard to DBP formation, the question is the contribution of SOC relative to DOC as precursors. There has been considerable attention directed at improving the processing of samples for DOC analyses (Wangersky 1990; Sharp and Peltzer 1993; Kaplan 1994), so separation of DOC from SOC should not be a hindrance for the industry. A very limited amount of research on organic particles contributing to DBP formation has been done with Aldrich humic acid (Najm et al. 2000). That material, however, is not representative of natural aquatic humic substances (Malcolm and McCarthy 1986) and is quite probably devoid of much protein, amine, or carbohydrate character that is associated with natural SOC. An additional aspect of the interim D/DBP Rule is the possibility that uv-specific absorbance may be included, and that will require filtration of samples for both absorbance and DOC.

OBJECTIVES AND RESEARCH GOALS

The principal objective of this research was to quantify analytical accuracy of TOC measurements. Three research goals followed this objective and constituted the major activity on this project. The first goal was (1) to develop TOC standards that can be used with all TOC methodologies to evaluate performance. Standards needed to include materials capable of assessing the ability to quantitatively separate IC from TOC and oxidize recalcitrant organic carbon. The second goal was (2) to determine the ability of different TOC methodologies to accurately measure TOC. This work included sample matrix considerations and involved numerous methodologies and instruments. The third goal was (3) to test the ruggedness of the proficiency standards in a round-robin survey involving a large number of laboratories. Most of these laboratories were in drinking water utilities, but some academic laboratories and U.S. government laboratories that work on drinking water issues and laboratories of five TOC manufacturers were also included.

A secondary, subordinate objective of the research was to quantify the analytical efficacy of TOC measurements. We addressed the question of whether TOC or DOC is the more appropriate measure of reactive NOM when the ultimate issue is regulation of DBP formation. One additional research goal followed from this objective. The fourth goal of the project was (4) to determine the contribution of natural organic particles, SOC, in the formation of DBP's. This was accomplished with tests on multiple pairs of filtered and unfiltered raw water samples.

CHAPTER 2
EXPERIMENTAL APPROACH

PROJECT COORDINATION AND RESEARCH PARTICIPANTS

The research project was coordinated through the Stroud Water Research Center. All standards were prepared at the Stroud Center and distributed to each of the project participants. Project participants were solicited initially by phone and then a letter with details of participant responsibilities to those expressing an interest. An attempt was made to engage laboratories that included drinking water utilities, academic laboratories, government laboratories, and TOC instrument manufacturers. Close collaboration with one government laboratory from the United States Geological Survey, Water Resources Division laboratory in Boulder, Colorado, under the direction of Dr. George Aiken, helped in the selection of materials to be screened for standards, contributed to the analysis of those materials, and helped identify what standards would be distributed to the other participants during the initial screening of candidate compounds. Arrangements were made with Huffman Laboratories, Inc. in Golden, Colorado, for the testing by elemental analysis of each of the reagents or stock materials in solid form that were used to prepare the standards. This provided a check on the chemical manufacturer, or in the absence of such information, provided organic carbon content. Philadelphia Suburban Water Company performed all of the disinfectant by-product formation potential testing associated with the fourth objective of this research.

SCREENING COMPOUNDS FOR TOC STANDARDS

We attempted to identify commercially available soluble compounds that could be used to verify oxidation efficiency in any TOC analyzer without bias towards a particular methodology. Based on previous experience, plus consultation of Standard Methods and a review of the literature, the following 24 compounds were considered as potential DOC standards: acetic acid, acetonitrile, antipyrine, 1-4 benzoquinone, beta cyclodextran, butanol, caffeine, dodecylbenzenesulfonic acid, ethylenediamine-tetraacetate (EDTA) as the disodium salt, fulvic acid, gluconic acid, glyoxal trimer dihydrate, humic acid, leucine, N-(2-thiazolyl)sulfanilamide, 1,10 phenanthrone, phthalic acid, polyacrylic acid (2,000 molecular weight), poly(styrene-sulfonic acid-co-maleic acid) [3:1 molar ratio], 2-propanol, tannic acid, tartaric acid, thiourea, and urea. Most of these compounds are commercially prepared and were purchased as reagent or higher grade. The dodecylbenzenesulfonic acid was only available in technical grade, and the humic and fulvic acid, isolated from the Suwannee River, were purchased from the International Humic Substances Society.

The compounds were dissolved in deionized water to achieve standard concentrations that were close to 2 mg C/L. Each compound was initially tested at the Stroud Center with three different TOC oxidation methodologies, (1) high temperature combustion (Shimadzu 5000 A), (2) uv-promoted persulfate (Sievers 800), and (3) heated, Pt-catalyzed persulfate (OI 700 and OI 1010). Following analysis at the Stroud Center, 40-mL vials of each dilute standard were sent to the USGS in Boulder for analysis with an OI 700.

Five different particle sources were screened to find suitable SOC standards, including C-18 beads, polystyrene polybeads, microcrystalline cellulose, freeze dried bacteria, and deciduous leaf powder. Waters Sep-Pak C-18 cartridges containing 55 to 105 µm sized silica sorbent [monofunctional silane chemistry, 12% carbon content] were the source of C-18 beads. A cartridge was cut open, the beads were removed, weighed, and then suspended in deionized water in a volumetric flask. Polystyrene polybeads (1.0 µm diameter) were purchased from Polysciences, Inc. and the microcrystaline cellulose (average particle diameter 2 to 20 µm) was Baker TLC reagent grade material. *Pseudomonas fluorescens* (strain P-17, American Type Culture Collection 49642) was transferred from a slant in the Stroud Center culture collection and grown on Kings B medium to confirm the fidelity of the working slant. Following confirmation, we inoculated 1 L of 0.1% tryptone yeast extract (TYE) and grew the organism to late log phase on a shaker table at room temperature. The cells were harvested by centrifugation and the resulting pellets resuspended in sterile phosphate buffer (pH 7). A total of three washes were performed to eliminate any TYE. The supernatant fluid from the final wash was decanted to waste and the pellet resuspended in cold deionized water. The suspensions were then freeze dried. Red maple leaves were collected after abscission, air dried, and ground in a Wiley mill with a Monel metal sieve of 60-mesh. The ground leaves were suspended in deionized water, wet-sieved through a Nytex sieve series to eliminate any particles greater than 50 µm in diameter and dried. Subsamples of all particles were sent to Huffman Laboratories for the determination of inorganic carbon and organic carbon by elemental analysis. The organic carbon content of the particles was used to prepare standards of suspensions with concentrations of approximately 2 mg C/L.

DELIVERY OF SUSPENDED PARTICLES TO TOC OXIDATION ZONE

We used direct microscopic counts to assess the ability to deliver P-17 from a well stirred suspension to the oxidation zone of two TOC analyzers, the reaction vessel of an OI 700 and the catalyst tube of a Shimadzu 5000 A equipped with a particle handling kit. A P-17 suspension (2 mg C/L) was subsampled in triplicate to provide an estimate of cell density in the suspension, and then the suspension was stirred and a sample collected by disconnecting the inlet tubing and intercepting the suspension delivered to either the reaction vessel or the catalyst tube. All samples were formalin fixed and prepared for direct microscopic counts using epifluorescence microscopy (Hobbie et al. 1977).

REPRODUCIBILITY AND STABILITY OF STANDARDS

The reproducibility and stability of different standards were assessed by preparing forty 40-mL vials of the standard, measuring TOC initially, and then after exposure to heat and light. All glassware was rendered organic-C free by heating to 500°C for 6 hours, and silicone-backed Teflon septa were cleaned in 5% w/v sodium persulfate heated to 60°C for 0.5 h and then rinsed 3 times with deionized water (Kaplan 1992). Standards were prepared in 5L volumetric flasks and dispensed to 40-mL vials with a peristaltic pump fitted with Pharmed and Teflon tubing. Particle suspensions were stirred constantly with a magnetic stir bar while they were dispensed. The vials were sealed with septa and screw caps and then the standards were Tyndallized by heating to 70°C for 0.5 h, allowed to cool to room temperature, and then heated in the same manner 24 h later. TOC was measured in 10 vials each, immediately after standards were dispensed to the vials, following the Tyndallization procedure, and then after 7 days of storage of the standards at room temperature in the light or in the dark. As a further check on the reproducibility and stability of standards, concentrations in 10 vials from each lot were measured in three TOC analyzers at the Stroud Center before any standards were sent to other laboratories, and then the measurements were repeated after the last data set was received from participants measuring standards from that lot.

MATRIX MODIFICATION

The influence of matrix modification on TOC measurements was assessed by amending standards with inorganic carbon from sodium bicarbonate (30 mg IC/L) to simulate increased alkalinity and mineral salts (conductivity of 512 µmhos) to simulate increased dissolved solid loading. The sodium carbonate was combusted at 550°C for 6 h to remove any potential organic residues and then tested for inorganic and organic carbon content. Mineral salts consisted of 10 mg/L CaCl2.2H2O, 50 mg/L MgSO4.7H2O, 10 mg/L KH2PO4, 10 mg/L Ca(NO3)2.4H2O, 1 mg/L Al2(SO4)3.18H2O, and 10 mg/L NH4Cl. Each salt was tested individually for the presence of organic carbon. The influence of matrix modification was assessed further through the method of additions. At the Stroud Center, samples from White Clay Creek were analyzed for TOC, with and without amendment with an organic carbon standard. The same procedure was carried out by utility participants, wherein treatment plant influent, raw water, was analyzed with and without amendment of an organic carbon standard provided by the Stroud Center.

EXCHANGE OF PROFICIENCY EVALUATION STANDARDS

Participants were sent standards in three shipments during the project. Each shipment consisted of a series of 8 standards, each supplied in triplicate 40-mL vials, and packed in a styrofoam shipper. The vials had labels with a standard number and a vial number, but the identity and concentration of the standard were not provided. Each participant received a data sheet for reporting, was given a code, and was asked to analyze and return the data within a 4- to 6-week period. Cover letters accompanied each shipment and provided some information to help the analyst, such as identifying the standards containing suspensions. One shipment included an analyst questionnaire (Appendix A) designed to reveal procedural aspects of each laboratory.

Standards prepared in a deionized water matrix were the focus during the first exchange, followed by a focus on inorganic carbon and mineral salt matrix modifications for subsequent sets (Table 2.1). With the second and third exchanges, the identity of the previous standards and the results from all laboratories were shared with the analysts. In addition to the difficult to oxidize solutions and the particle suspensions, we always included a potassium hydrogen phthalate (KHP) standard to assess how well the detector for each instrument was calibrated. Solutions of KHP amended with a 30 mg C/L inorganic carbon spike were sent out twice to quantify the ability of each analyst to separate inorganic from organic carbon, and a standard consisting low-OC deionized water was sent out once to assess the condition of the instrument and laboratory environment. One difficult to oxidize organic compound, dodecylbenzenesulfonic acid, was included with each shipment. This allowed us to generate a larger database on this potential standard and assess whether an operator was generating consistent data over all three exchanges. Some analysts requested more than one set of standards so that they could analyze the samples with different model instruments.

Table 2.1
Proficiency evaluation standards prepared for participating laboratories

Shipment number	Standard	Identity	Preparation date	Concentration (mgC/L)
1	1	Caffeine	8-13-98	2.031
	2	Dodecylbenzene sulfonic acid	8-17-98	1.983
	3	Humic acid	8-18-98	1.699
	4	Polystyrene polybeads	8-19-98	1.976
	5	Potassium hydrogen phthalate	8-20-98	2.009
	6	Potassium hydrogen phthalate w/$NaHCO_3$	8-20-98	2.009
	7	Deionized water	8-12-98	-----
	8	P17	8-31-98	2.003
2	9	Potassium hydrogen phthalate	2-02-99	4.003
	10	Potassium hydrogen phthalate w/mineral salts	2-11-99	4.003
	11	Polystyrene polybeads	2-04-99	1.490
	12	Polystyrene polybeads w/mineral salts	2-18-99	1.515
	13	Dodecylbenzene sulfonic acid	2-08-99	2.803
	14	Fulvic acid w/mineral salts	2-16-99	2.742
	15	White Clay Creek storm flow	2-23-99	4.513
	16	Leaf powder	3-9-99	4.081
3	17	Caffeine	5-04-99	3.000
	18	Caffeine w/mineral salts & Na_2CO_3	6-03-99	3.000
	19	Potassium hydrogen phthalate	6-02-99	2.500
	20	Potassium hydrogen phthalate w/Na_2CO_3	6-07-99	2.500
	21	Dodecylbenzene sulfonic acid	5-04-99	3.502
	22	P17 w/mineral salts	6-15-99	2.698
	23	Humic/fulvic acid	6-08-99	3.678
	24	Humic/fulvic acid w/Na_2CO_3	6-08-99	3.678
	25	Dodecylbenzene sulfonic acid concentrate	6-09-99	186.9

INFLUENCE OF PARTICLES ON DISINFECTION BYPRODUCT FORMATION POTENTIAL (DBPFP)

Philadelphia Suburban Water Company (PSWC) developed a protocol for DBPFP tests and applied this protocol to samples from two different water sources (Crum Creek and Neshaminy Creek) under two different treatments, unfiltered and filtered through precombusted glass fiber filters (Whatman GF/F). Crum Creek has water drawn from a reservoir within a watershed dominated by agricultural uses. The Neshaminy Creek water comes from a watershed dominated by residential and industrial land uses, and receives wastewater treatment discharges. PSWC personnel collected the water samples and Stroud Center personnel filtered the water. Samples were taken for the DBPFP test, the measurement of TOC, ash free dry mass (AFDM), and chlorophyll a. The test protocol developed by PSWC began with an estimation of the chlorine demand of the filtered or unfiltered sample. A one-gallon sample of test water was adjusted to pH 8.0 with an excess of phosphate buffer and distributed into 10 bottles. A chlorine solution was added to each of the bottles in 0.5 mg/L increments. All bottles were topped off with the pH adjusted test water, capped, and incubated in the dark at 21°C for 24 hours. After incubation the final pH in each bottle was measured as well as the free and total chlorine residual to assess the breakpoint chlorination level. Two bottles were chosen for DBP testing, one that had a chlorine dose close to the operating applied dose, and one that was 0.5 mg/L above the breakpoint.

The test thus resulted in a characterization of the chlorine demand of the water samples and a determination of DBP's for the waters at two applied chlorine doses at uniform incubation conditions of time, temperature, and pH. Each sample was analyzed at PSWC for the following DPB's: (1) Haloacetic Acids (monochloroacetic acid, monobromoacetic acid, dichloroacetic acid, dibromoacetic acid, trichloroacetic acid, and bromochloroacetic acid); (2) Pentane extractable compounds (a) Trihalomethanes (THMs) (chloroform, bromoform, bromodichloromethane, and dibromochloromethane, (b) Haloacetonitriles (HANs) (trichloroacetionitrile, dichloroacetonitrile, dibromoacetonitrile, and bromochloroacetonitrile), (c) Three other compounds (dichloroacetone, trichloroacetone, and chloropicrin); and (3) Chloral hydrate.

Samples of filtered and unfiltered water were analyzed at the Stroud Center. Filtrates were analyzed for DOC with an OI 700 and Shimadzu 5000 A, while unfiltered water was analyzed for TOC with the Shimadzu 5000 A, total suspended solids (Standard Methods 2540 D), and fixed and volatile solids (Standard Methods 2540 E). Chlorophyll a was analyzed at PSWC on the unfiltered samples (Standard Methods 10200 H).

STATISTICAL ANALYSES

Data from all participants in the study were placed into SAS data sets (SAS Institute, Cary NC) maintained on a microVAX 3100 at the Stroud Center. Statistical analyses included t-test, analysis of variance (ANOVA), and univariate analyses. Significant differences were determined with an alpha error level of $P = 0.05$. Relative precision was defined as the coefficient of variation for replicate analyses, expressed as a percent, and recovery was defined as the ratio of the measured concentration to theoretical concentration, expressed as a percent. In all of the histograms in this report, the values presented are the mean, and when error bars are included, they represent one standard deviation.

The hierarchal nature of the data set of TOC analyses, involving 3 different TOC methods, 12 different instrument models, and 39 different machines or analysts, lent itself to statistical testing with a nested analysis of variance (ANOVA). The nested ANOVA was used to assess the error associated within each level of this analytical hierarchy. The first step in this analysis was to calculate the relative analytical error (RAE) for each standard analyzed. RAE is defined as the absolute difference between theoretical and measured concentrations, divided by the theoretical concentration. The error term for each analysis was derived from the mean square of the adjacent and lower analytical level. Thus, to assess the method effect, the model nested instrument within method, and the mean square of the instrument from the general linear models procedure was used as the error term. Similarly, to assess the instrument effect, machine was nested within instrument and the machine mean square was the error term. Lastly, for the machine effect, there was no nesting, and the mean square of replicate measurements was the error term. Following the ANOVA, a Tukey's Studentized Range Test was run to identify which means were significantly different. There were a few instances where the ANOVA did not reveal significance, but the Tukey's did. We only considered those differences judged significant by both tests as statistically valid.

CHAPTER 3
RESULTS AND DISCUSSION

PARTICIPATING LABORATORIES AND TOC ANALYZERS

A total of 33 laboratories and 39 TOC analyzers were involved in this research project. Participants included 19 water utility laboratories, 5 academic laboratories, 2 government laboratories, and 2 private laboratories, and 5 TOC instrument manufacturers (Table 3.1). The TOC analyzers represent 12 different models and 6 manufacturers, with high temperature combustion and uv-promoted persulfate the major oxidation types (Table 3.2). While no potential participants for the study were turned away, several utilities felt unable to devote the time needed to participate in this research.

Table 3.1
Participants analyzing TOC standards

Participant	Location
Water Utilities	
AWW Service Co., Inc.	Belleville, IL
Belleville Lab	
Buffalo Pound Water Treatment Plant	Moose Jaw, SA
Charleston Commissioners of Public Works	Charleston, SC
City of Chesapeake	Chesapeake, VA
Department of Public Utilities	
City of Cincinnati	Cincinnati, OH
Department of Water Works	
City of Houston	Houston, TX
Department of Public Works	
Consumers Pennsylvania Water Company	Sharon, PA
Shenango Valley Division	
Greater Vancouver Regional Water District	Burnaby, BC
Metropolitan Utilities District	Omaha, NE
North Jersey District Water Supply Commission	Wanague, NJ
North Penn Water Authority Laboratory	Chalfont, PA
Philadelphia Suburban Water Company	Bryn Mawr, PA
Philadelphia Water Department	Philadelphia, PA
Bureau of Laboratory Services	
Portland Water Bureau	Portland, OR
South Central Connecticut Regional Water Authority	New Haven, CT
Tennessee-American Water Company	Chattanooga, TN
Tampa Water Department	Tampa, FL
Water Quality Bureau	
San Francisco Public Utilities Commission	Millbrae, CA
Water Quality Laboratory	
Jefferson Parish	Harahan, LA

(continued)

Table 3.1 (continued)

Participant	Location
Academic Laboratories	
Center for Biofilm Engineering College of Engineering Montana State University – Bozeman	Bozeman, MT
Civil, Environmental & Architectural Engineering University of Colorado	Boulder, CO
Department of Civil Engineering University of Waterloo	Waterloo, ON
Ecole Polytechnique Department of Civil Engineering University of Montreal	Montreal, QU
Northwestern University Technological Institute	Evanston, IL
Government Laboratories	
Treatment Technology Evaluation Branch Water Supply and Water Resources Division U.S. EPA National Risk Management Research Laboratory	Cincinnati, OH
United States Geological Survey Water Resources Division	Boulder, CO
Private Laboratories	
Huffman Laboratories, Inc.	Golden, CO
Stroud Water Research Center	Avondale, PA
TOC Instrument Manufacturers	
Anatel Corporation	Boulder, CO
OI Analytical	College Station, TX
Shimadzu Scientific Instruments, Inc.	Columbia, MD
Sievers Instrument, Inc.	Boulder, CO
Tekmar-Dohrmann	Cincinnati, OH

Table 3.2
TOC instruments in participating laboratories

Instrument models	IC removal	Method of oxidation	Detection	Number of machines
Anatel A-2000	Internal sparge	UV-promoted wet oxidation	NDIR*	1
Astro 2001	Internal sparge	UV-promoted wet oxidation	NDIR	2
Dohrmann Apollo 9000	External sparge	High temperature combustion	NDIR	1
Dohrmann DC 80	External sparge	UV-promoted wet oxidation	NDIR	1
Dohrmann DC 180	External sparge	UV-promoted wet oxidation	NDIR	5
Dohrmann Phoenix 8000	External sparge	UV-promoted wet oxidation	NDIR	6
OI 700	Internal sparge	Pt-catalyzed wet oxidation	NDIR	5
OI 1010	Internal sparge	Pt-catalyzed wet oxidation	NDIR	2
OI 1020	External sparge	High temperature combustion	NDIR	1
Shimadzu 5000A	External sparge	High temperature combustion	NDIR	9
Shimadzu 5050	External sparge	High temperature combustion	NDIR	2
Sievers 800	Internal sparge	UV-promoted wet oxidation	Conductivity	4

* NDIR = nondispersive infra-red detector

SCREENING COMPOUNDS FOR TOC STANDARDS

Six of the test compounds, acetonitrile, caffeine, dodecylbenzenesulfonic acid, fulvic acid, humic acid, and polystyrene-sulfonic acid proved difficult to oxidize completely in all instruments tested at the Stroud Center and USGS (Fig. 3.1). This was assessed by comparing the empirical concentrations measured in each instrument to the theoretical concentration for each standard. In addition, altering the oxidant level in the Sievers 800 while analyzing caffeine tested the potential of that compound to reveal reduced recoveries under suboptimal conditions. We began by measuring caffeine in a deionized water matrix and had an 88% oxidation efficiency when compared to our theoretical value (1.80 ± 0.02 mg C/L measured and 2.04 mg C/L expected). Increasing the rate of oxidant injection during the analysis from 0.75 μl/min to 1.00 μl/min did not improve recovery (1.78 ± 0.01 mgC/L), but decreasing the oxidant flow to 0.25 μl/min reduced recovery to 83% (1.70 ± 0.05 mg C/L), indicating that caffeine could reveal decreased oxidation recovery when an instrument is not operated optimally.

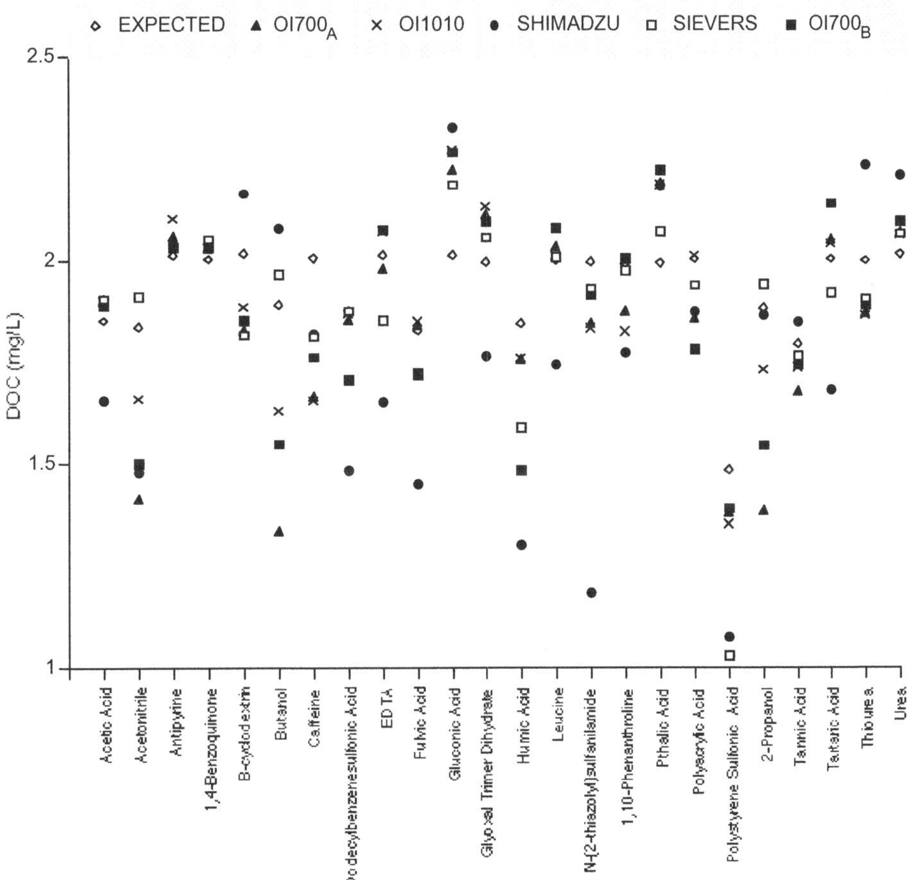

Figure 3.1 Initial screening of compounds to select proficiency evaluation standards

Two potential standard compounds, copper phthalocyanine used as a British Standard (British Standards Institution 1997), and pyridine cited in Standard Methods, were eliminated from consideration because they are toxic. The alcohols tested also were unacceptable because they introduced machine bias. Specifically, butanol and propanol were underestimated by all methods that heat the sample during the sparging step designed to remove inorganic carbon. In fact, the instrument with the highest temperature setting, the OI 700, had the lowest recovery. Other instrument-to-instrument differences such as those observed with acetate, EDTA, B-cyclodextrin, and thiourea are less easily explained. Statistical analyses of these results included a test of the hypothesis that the difference between the theoretical and empirical values was equal to zero. For the six compounds indicated above, the differences were statistically significant at P-value of 0.05 or less.

All of the particles screened presented recovery problems for the TOC analyzers, with a maximum recovery of 77% and a minimum recovery of 4% (Table 3.3). The organic C content of these materials, determined by Huffman Laboratories, was 91.3% for the polybeads, 41.8% for P-17, and 44.1% for cellulose. We were unable to disperse the C-18 material because of its extreme hydrophobic nature. The C-18 beads accumulated at the meniscus in the volumetric flask and could not be dispersed to a uniform suspension. The microcrystalline cellulose particles, at 2-to 20 μm diameter, were smaller than the 20 to 100 μm particles recommended in the British standard for TOC analysis (British Standards Institution 1997), but still showed visible settling within the sample tubes prior to analysis. The other particles also settle, but only over a more extended storage period. Some preliminary tests were carried out with the cellulose, but eventually both of these materials were eliminated from consideration.

Table 3.3
Measurement of suspended organic carbon standards

		Concentration (mg C/L)			
			Measured ($\bar{x}$ ± SD (n)) [% recovery]		
Compound	Date	Expected	1010*	5000A*	800*
Polystyrene Polybeads	23 Jul 98	2.052	0.201 ± 0.016 (4) [9.8]	1.579 ± 0.047 (5) [77.0]	0.736 ± 0.038 (2) [35.8]
P-17	7 May 98	1.995	1.054 ± 0.055 (2) [52.8]	0.480 ± 0.053 (5) [24.1]	1.425 ± 0.004 (2) [71.4]
Cellulose	12 May 98	2.017	0.115 ± 0.005 (2) [5.7]	1.317 ± 0.559 (5) [65.3]	0.076 ± 0.002 (2) [3.8]

*TOC analyzers used include an OI 1010, Shimadzu 5000A, and a Sievers 800

DELIVERY OF SUSPENDED PARTICLES TO TOC OXIDATION ZONE

The low and variable recovery of particles in the TOC analyzers may involve several factors in addition to oxidation efficiency. With any suspension, the problems of obtaining a uniform, representative sample and delivering that sample to the oxidation zone are the first obstacles to overcome. To eliminate settling or heterogeneity of the suspension as factors, suspensions of P-17 were sampled while continuously stirred. A 200 mg C/L stock of P-17 was subsampled and each subsample was used to prepare 100-fold dilutions. The four resulting 2 mg C/L suspensions in volumetric flasks were stirred continuously while they were sampled as grab samples with a sipper on an OI 700. The measured concentrations for 10 replicate measurements from these solutions were 1.482 ± 0.089, 1.491 ± 0.104, 1.460 ± 0.140, and 1.451 ± 0.145 mg

C/L ($\bar{x}$ ± SD). These data demonstrate three things: (1) dilutions from a concentrated suspension can be prepared with reasonable precision, as the average coefficient of variation for the four suspensions was 1.3%; (2) the analytical precision for the 10 replicate determinations from each suspension was reasonable (range of 6 to 10%) but 3- to 10-fold higher than the analytical precision for a true solution (Kaplan 1994); (3) the recovery efficiency for the suspensions, under conditions considered optimal for keeping the particles in a uniform suspension, was 73 to 75%.

We used direct microscopic counts to assess the ability to deliver P-17 from a well-mixed suspension to the reaction vessel of an OI 700 and the combustion zone of a Shimadzu 5000 A. The 2 mg C/L P-17 standard, a suspension, had a cell density of 1.21 ± 0.12 x 107 cells/mL. The suspensions collected just prior to the oxidation zone of the OI 700 had 7.32 ± 1.32 x 106 cells/mL, representing a 40% reduction, and 7.10 ± 1.32 x 106 cells/ mL in the Shimadzu 5000A, representing a 41% reduction in P-17 prior to oxidation. Earlier preliminary work with the OI 700 had shown similar losses after sampling 10 mL of suspension (two cycles of the 5.0 mL sample loop), but also showed an increase in cell concentration after another 20 mL were collected, demonstrating that association and disassociation of the bacterial cells with the tubing surfaces has a potentially significant, though variable influence of the delivery of suspended particles to the oxidation zone. Additionally, the analysis of deionized water blanks in the OI 700 was unstable for several analytical cycles following the analysis of P-17, indicating a delay in moving all the cells through the instrument.

REPRODUCIBILITY AND STABILITY OF STANDARDS

Standard materials prepared for an interlaboratory exchange require a high degree of precision, that is, there should be minimal vial-to-vial variation within a batch of standard. Within batch variation can be related to the preparation of the vials and septa used for each standard, and the care taken in dispensing solutions and suspensions to individual vials as well as the handling of the vials during the Tyndallization step and storage. The average coefficient of variation for all batches of solutions, based on analysis of 10 replicate vials for each standard in an OI 700, was 1.55 ± 1.17% (n=22). Reproducibility with the suspensions used as standards was more difficult to assess, because of the problems associated with analysis of SOC. Nevertheless, the average coefficient of variation with suspensions analyzed in an OI 700 was 11.2 ± 5.9% (n=8).

The stability of different standards when exposed to heat and light are important criteria for several reasons. First, Tyndallization helps prevent bacterial growth in the standards that are susceptible to biodegradation, and all standards were heat treated prior to distribution to participants. Second, the handling of standards, once received by participating laboratories, was difficult to standardize, so while some participants may have kept the standards in the refrigerator, others may have kept them on the bench top. Lastly, any standard should have a shelf life of at least one week, if not considerably longer, to give the analyst flexibility in accomplishing the work with the standard. Caffeine, acetonitrile, humic acid, and P-17 were stable to both heat and light, changing by a maximum of 3% from the different treatments when measured with an OI 700 (Table 3.4). The work with P-17 was subjected to extended storage in the dark, but with less replication. We found decreases of 20% with concentrations of 1.193 ± 0.243 after 24 days, and 1.184 ± 0.159 after 31 days, each based on duplicate vials.

Table 3.4
Influence of heat and light on stability of standards

DOC Concentration (mg C/L)

[$\bar{x}$ ± SD, n = 10]

Standard	Initial	Tyndallized	7-d Light	7-d Dark
Caffeine	1.726 ± 0.005	1.749 ± 0.022	1.751 ± 0.016	1.740 ± 0.008
Acetonitrile	1.918 ± 0.017	1.970 ± 0.038	1.910 ± 0.017	1.907 ± 0.008
humic acid	1.578 ± 0.024	1.575 ± 0.009	1.565 ± 0.007	1.575 ± 0.013
P-17	1.477 ± 0.115	1.482 ± 0.073	1.440 ± 0.104	1.481 ± 0.131

A much more extensive test of standard stability involved measurements of concentrations before each batch of standards was sent out to the participants, and measurements after all participants had finished their analyses. This was done for each standard. The percent change for all solutions ranged from 0.19 to 6.75% over a 57 to 113 day period. For the particles, the range was 0.56 to 36.90% over a 49 to 99 day period.

MATRIX MODIFICATION

The roles of the inorganic and organic matrices in influencing TOC analytical precision and accuracy needs to be considered because the natural waters analyzed by water utilities for TOC have complex matrices that vary within and between source waters. A caffeine standard prepared in deionized water and measured with a Sievers 800 gave a concentration of 1.80 ± 0.02 mg C/L. When a 200 mg IC/L solution of sodium bicarbonate was added the standard measured 1.77 ± 0.01 mg C/L and when a mineral salts solution with a conductivity of 534 μSiemens was added the standard measured 1.80 ± 0.11 mg C/L. The mineral salts data obviously showed no influence on recovery, and the alkalinity adjustment resulted in only a 1.5% decrease in recovery.

A more robust test of matrix influences involves the amendment of a natural sample with a standard solution (method of additions). Samples from White Clay Creek were collected during a storm and filtered to remove particles. When DOC was measured in 10 of the samples with and without the addition of caffeine, we observed a decline in recovery relative to the 88% recovery of caffeine dissolved in deionized water, ranging from 0 to about 6% (Table 3.5). The decrease in recovery is not strictly related to DOC concentration or conductivity, as the sample from 1330 h had high DOC, moderate conductivity, and relatively high recovery, but all other samples with high DOC concentrations had diminished recovery.

Table 3.5
Influence of a complex natural matrix on recovery of caffeine with a Sievers 800 TOC analyzer

		DOC (mgC/L [$\bar{x}$ ± SD, n = 2])			
Date	Time	Streamwater	Streamwater + caffeine	Caffeine [% recovery]	Conductivity (μSiemens)
23JAN98	0530	0.944 ± 0.04	2.74 ± 0.10	1.80 ± 0.11 [88.24]	220
	1030	1.85 ± 0.03	3.65 ± 0.09	1.80 ± 0.10 [88.24]	204
	1230	3.91 ± 0.02	5.63 ± 0.04	1.72 ± 0.04 [84.31]	205
	1330	4.82 ± 0.04	6.62 ± 0.02	1.80 ± 0.04 [88.24]	175
	1530	5.86 ± 0.01	7.58 ± 0.03	1.72 ± 0.03 [84.31]	112
	1730	7.00 ± 0.03	8.69 ± 0.01	1.69 ± 0.03 [82.84]	90
	1930	7.55 ± 0.04	9.23 ± 0.01	1.68 ± 0.04 [82.35]	103
	2330	7.18 ± 0.04	8.92 ± 0.01	1.74 ± 0.04 [85.29]	128
24JAN98	0330	5.90 ± 0.04	7.63 ± 0.06	1.73 ± 0.07 [84.80]	152
	0830	4.59 ± 0.03	6.32 ± 0.01	1.73 ± 0.03 [85.22]	172

The influence of matrix modification on TOC measurements was also tested using acetonitrile (2 mg C/L), a sodium bicarbonate (30 mg IC/L) amendment to simulate increased alkalinity, mineral salts (conductivity of 512 μmhos) to simulate increased dissolved solid loading, and storm samples from White Clay Creek to simulate loading of complex organic carbon. The studies were carried out with three different instruments (OI 1010, Sievers 800, and Shimadzu 5000 A). The bicarbonate amendments influenced all instruments nearly equally, but mineral salts interfered with the two persulfate methods while having only a slight effect on the high temperature instrument (Table 3.6). Matrix modification due to the influence of storm waters showed little influence on the persulfate methods, with all recoveries exceeding 90%, and most exceeding 95% for the OI 1010 (Table 3.7) and 93% for the Sievers 800 (Table 3.8). The high temperature system, however, had recoveries in the 75 to 85% range (Table 3.9). The storm water matrix effect, at least for storm waters from White Clay Creek, is more of an organic influence than an inorganic influence. Conductivity decreases during storms as the proportionate contribution to stream flow of low conductivity rainwater increases. It is possible that the acetonitrile reacts with some of the hydrophobic DOC present in storm water and is lost to the sample vial during the acidification and external sparging step associated with the high temperature system. Acidification of surface waters followed by sparging has been shown to result in measurable, concentration-dependent losses of DOC (Kaplan 2000).

Table 3.6
Influence of matrix modifications on acetonitrile recovery

	DOC Concentration (mg C/L [$\bar{x}$ ± SD, n = 2]) [% recovery]		
Analyzer	Unamended	Bicarbonate	Mineral Salts
OI 10102.052 ± 0.007	1.920 ± 0.037	1.570 ± 0.022 [93.5]	[76.5]
Sievers 800	2.033 ± 0.046	1.942 ± 0.076 [95.6]	1.683 ± 0.012 [82.8]
Shimadzu 5000A	1.986 ± 0.010	1.906 ± 0.048 [96.0]	1.902 ± 0.057 [95.7]

Table 3.7
Recovery of acetonitrile added to storm samples and analyzed with an OI 1010

		DOC Concentration (mg C/L) [$\bar{x}$ ± SD (n = 2)]		
Storm Sample	Date	Streamwater	Streamwater + acetonitrile	recovery (%)
1	11 May 98	8.596 ± 0.152	10.700 ± 0.006	102.6
2	11 May 98	6.302 ± 0.101	8.303 ± 0.004	97.5
3	12 May 98	5.662 ± 0.069	7.680 ± 0.004	98.3
4	12 May 98	7.442 ± 0.014	9.305 ± 0.074	90.8
5	12 May 98	5.746 ± 0.057	7.703 ± 0.057	95.4
6	12 May 98	6.964 ± 0.002	8.956 ± 0.007	97.0
7	12 May 98	7.464 ± 0.016	9.607 ± 0.056	104.4
8	12 May 98	5.462 ± 0.045	7.438 ± 0.065	96.3
9	13 May 98	4.067 ± 0.025	6.042 ± 0.010	96.2
10	13 May 98	3.377 ± 0.010	5.382 ± 0.069	97.7

Table 3.8
Recovery of acetonitrile added to storm samples and analyzed with a Sievers 800

Storm Sample	Date	Streamwater	Streamwater + acetonitrile	Recovery (%)
		DOC Concentration (mg C/L) [$\bar{x}$ ± SD (n = 2)]		
1	11 May 98	7.671 ± 0.111	9.533 ± 0.071	91.6
2	11 May 98	5.760 ± 0.025	7.696 ± 0.069	95.2
3	12 May 98	5.278 ± 0.000	7.273 ± 0.039	98.1
4	12 May 98	7.223 ± 0.014	9.086 ± 0.055	91.7
5	12 May 98	5.760 ± 0.000	7.685 ± 0.014	94.7
6	12 May 98	6.928 ± 0.039	8.871 ± 0.005	95.6
7	12 May 98	6.760 ± 0.018	8.643 ± 0.021	92.6
8	12 May 98	5.051 ± 0.034	6.958 ± 0.007	93.8
9	13 May 98	3.803 ± 0.032	5.775 ± 0.025	97.0
10	13 May 98	2.960 ± 0.145	5.029 ± 0.069	101.6

Table 3.9
Recovery of acetonitrile added to storm samples and analyzed with a Shimadzu 5000A

Storm Sample	Date	Streamwater	Streamwater + acetonitrile	recovery (%)
		DOC Concentration (mg C/L) [$\bar{x}$ ± SD (n = 2)]		
1	11 May 98	6.395 ± 0.245	8.319 ± 0.478	96.9
2	11 May 98	5.096 ± 0.124	6.637 ± 0.111	77.6
3	12 May 98	4.733 ± 0.140	6.352 ± 0.194	81.5
4	12 May 98	6.527 ± 0.269	8.139 ± 0.157	81.2
5	12 May 98	4.990 ± 0.030	6.747 ± 0.009	88.4
6	12 May 98	6.331 ± 0.312	7.827 ± 0.354	75.4
7	12 May 98	6.181 ± 0.158	7.870 ± 0.288	85.0
8	12 May 98	4.679 ± 0.098	6.364 ± 0.026	84.8
9	13 May 98	3.426 ± 0.091	5.233 ± 0.022	91.0
10	13 May 98	2.959 ± 0.028	4.570 ± 0.061	81.1

EXCHANGE OF PROFICIENCY EVALUATION STANDARDS

Data from the laboratories are reported below within a framework of four general categories, (1) easily oxidized organic molecules, (2) difficult to oxidize organic molecules, (3) inorganic matrix influences on measurements of organic molecules, and (4) analysis of organic particles. In general, all laboratories performed well in the analysis of KHP, most performed well in the analyses of difficult to oxidize solutions of organic molecules, some laboratories had problems separating inorganic carbon from organic carbon, and all laboratories had difficulty with standards containing suspensions (Fig. 3.2). A useful context for the TOC evaluation is the

U.S. EPA certification range of ± 20% accuracy. The average recovery for the 39 instruments used to measure KHP was 103 ± 7%, with a range of 86 to 125%. All but one of the analysts would have qualified under the U.S. EPA guidelines of ± 20% accuracy, with the next lowest within ±14%. Thirty-three of the analyzers (85% of the total) were operated within 10% of the theoretical value, with 23 of them within 5% or better. When separated by the 12 instrument models, the range of recoveries was 96 to 113% (Fig. 3.3).

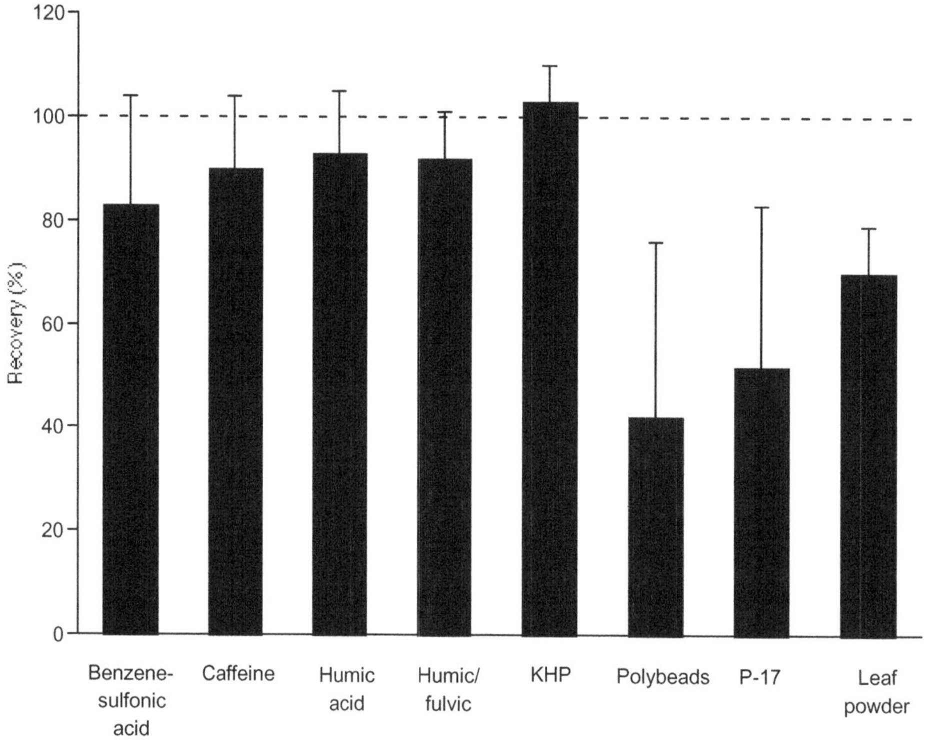

Figure 3.2 Average performance of all laboratories with proficiency evaluation standards (error bars represent S.D.)

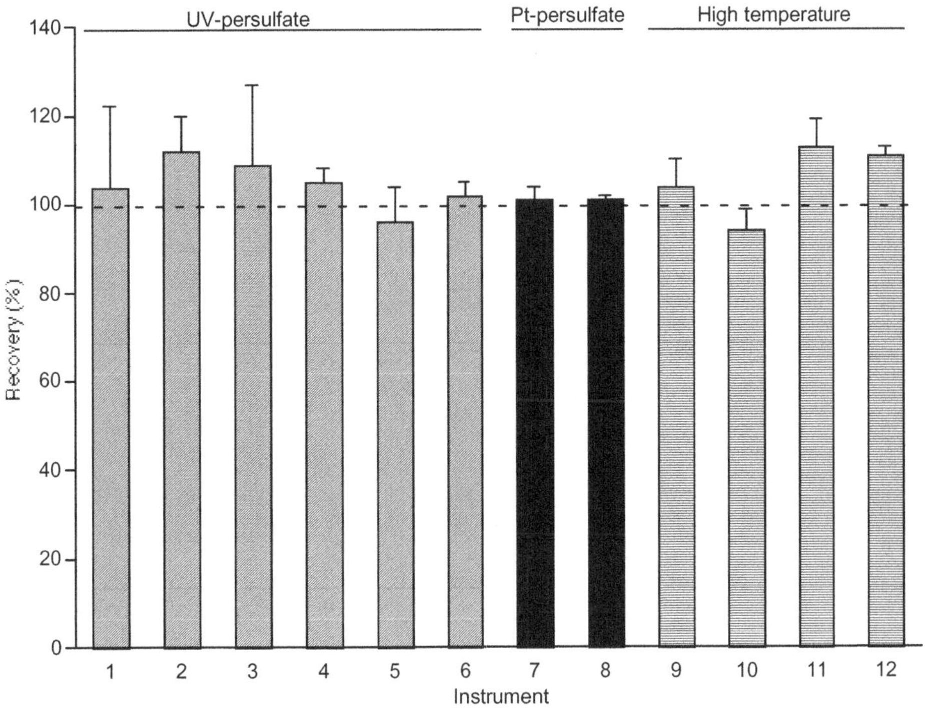

Figure 3.3 Recovery of KHP separated by TOC instrument

There was little influence of mineral salts on the analysis of KHP, as the average recovery for the 38 instruments involved was 102 ± 6%, with a range of 89 to 119%, and 92% of all instruments operated within 10% or better of the theoretical value. In contrast, several analysts had difficulty in efficiently separating an inorganic C amendment from the KHP standard when a 30 mg IC/L spike as added to the standard. With a 30 mg IC/L spike in a 2.5 mg C/L KHP standard, a removal efficiency of 98.3% would be required to avoid overestimating the true KHP concentration by 20% and thus fail U.S. EPA certification. Twelve of the 39 analysts would have failed based on that criterion. In addition to KHP, two other standards, caffeine and a humic/fulvic acid mix were amended with inorganic carbon, and seven analysts that worked with those amended standards would not have passed certification based on incomplete IC removal. Averaged across all analysts, the IC amendment of KHP had the greatest impact on recovery (Fig. 3.4). When separated by instrument model, several were not operated properly to adequately remove IC from KHP (Fig. 3.5), caffeine (Fig. 3.6), or the humic/fulvic acid mix (Fig. 3.7).

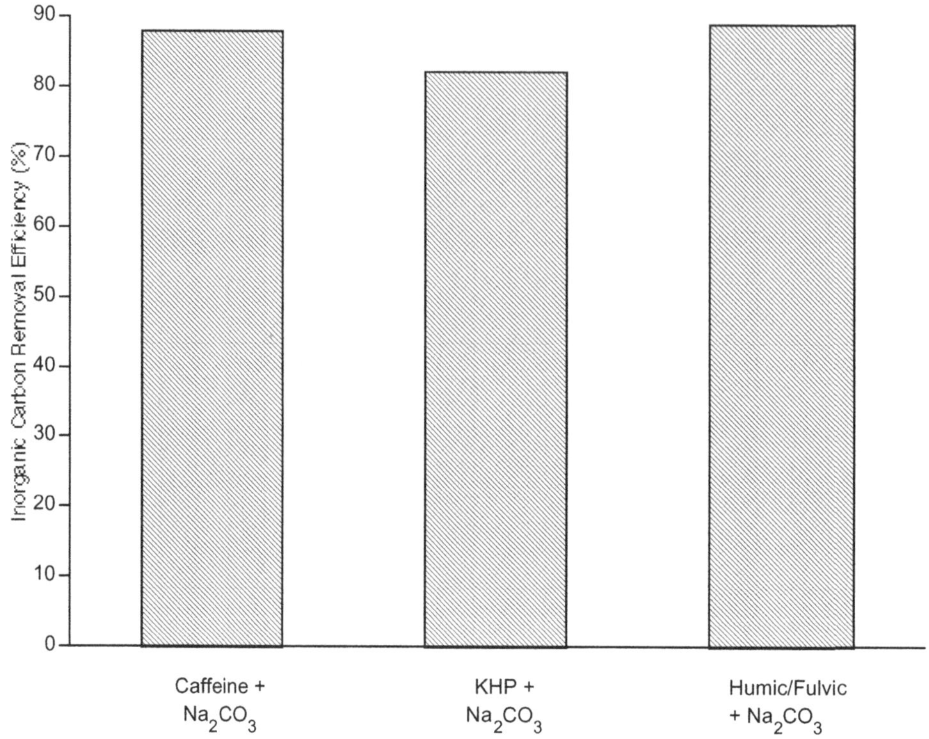

Figure 3.4 Influence of inorganic carbon amendments on TOC accuracy

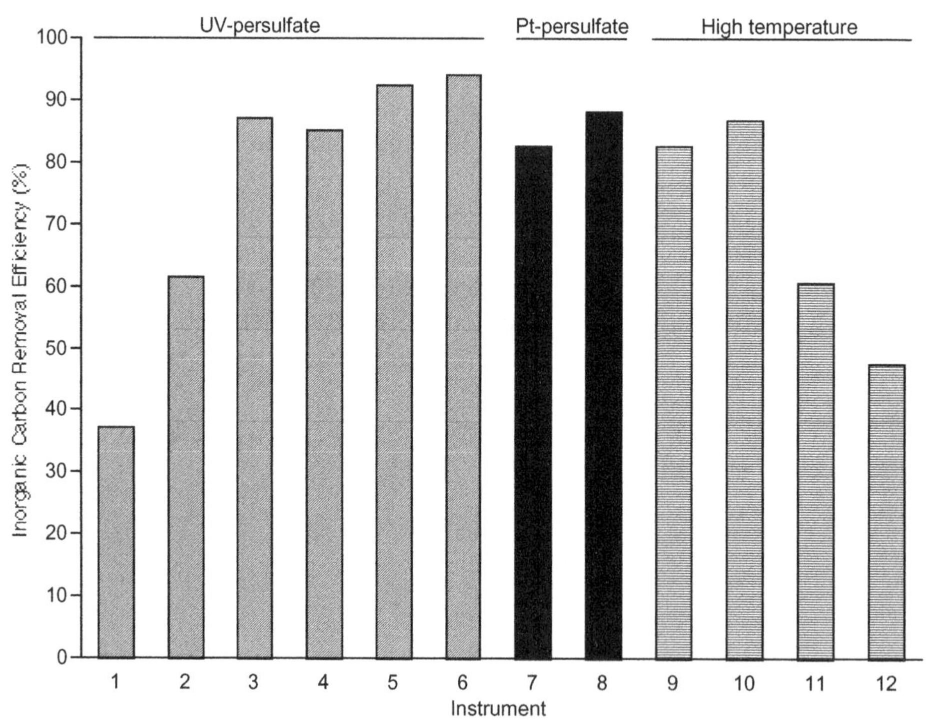

Figure 3.5 Influence of inorganic carbon amendments on TOC accuracy with KHP separated by TOC instrument

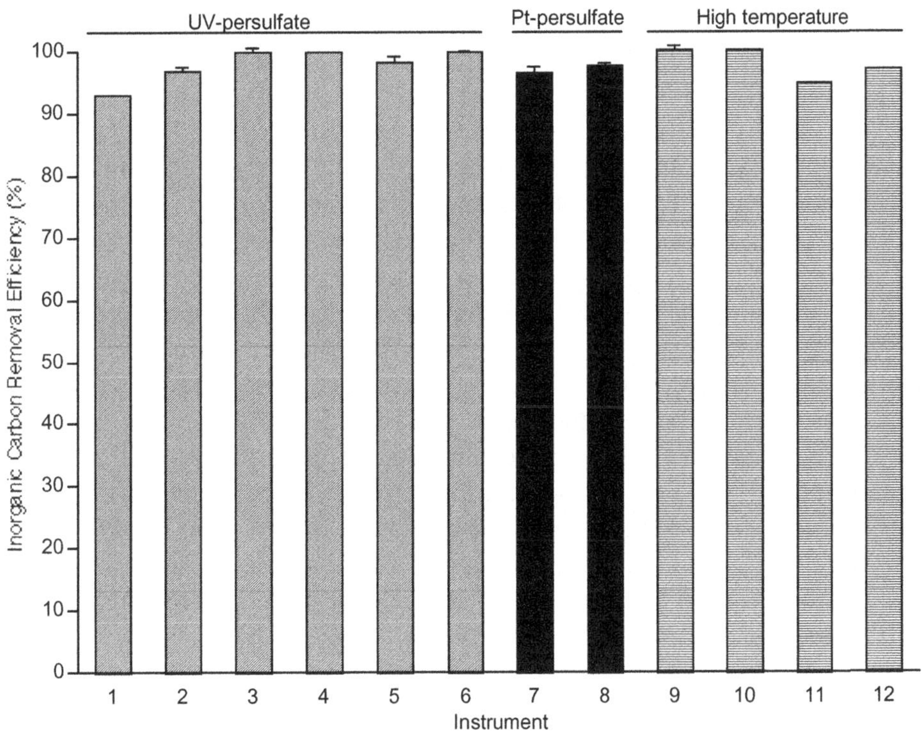

Figure 3.6 Influence of inorganic carbon amendments on TOC accuracy with caffeine separated by TOC instrument

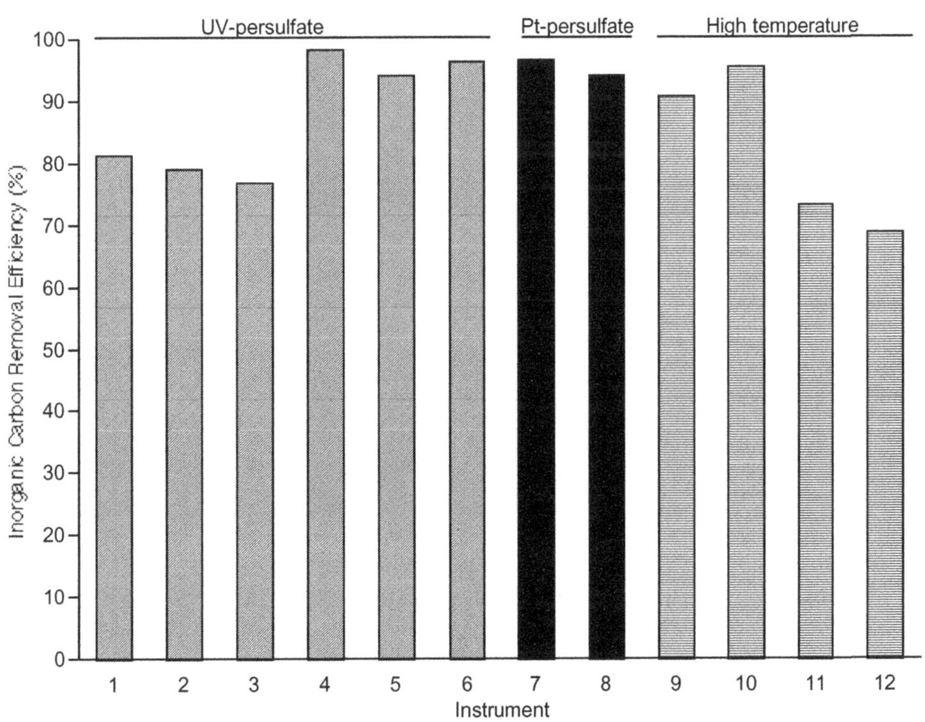

Figure 3.7 Influence of inorganic carbon amendments on TOC accuracy with humic/fulvic acid mix separated by TOC instrument

Dodecylbenzene sulfonic acid and caffeine were particularly difficult molecules for numerous analysts. For dodecylbenzene, 15 of the 39 analysts achieved less than 80% recovery. The average recovery for all analysts was 83 ± 23%, and the range was 43 to 109%. In separating the analyses by instrument model, some, though clearly not all, of the uv-persulfate instruments had the most difficulty with this standard, and 2 of the high temperature instruments overestimated the concentration (Fig. 3.8). However, closer inspection of the data suggests that the problems were more operator-associated. For example, the range of analytical accuracies over identical instrument models was 44 to 71% for the 6 Phoenix 8000 instruments, 52 to 73% for the 6 DC-180 instruments, and 43 to 95% for the 9 high-temperature Shimadzu 5000A instruments used.

Nine of 39 analysts achieved less than 80% recovery with caffeine. The average recovery for all analysts was 89 ± 14%, and the range was 63 to 114%. On average, all of the high temperature systems overestimated the concentrations of caffeine, and all of the uv-persulfate systems and the Pt-persulfate systems underestimated the caffeine concentrations, contributing to a reduction in recovery (Fig. 3.9). Ranges for operators of the same model instruments were slightly lower than those associated with analysis of dodecylbenzene sulfonic acid, including 63 to 80% for the 6 Phoenix 8000 instruments, 75 to 92% for the 6 DC-180 instruments, and 92 to 114% for the 9 Shimadzu 5000A instruments.

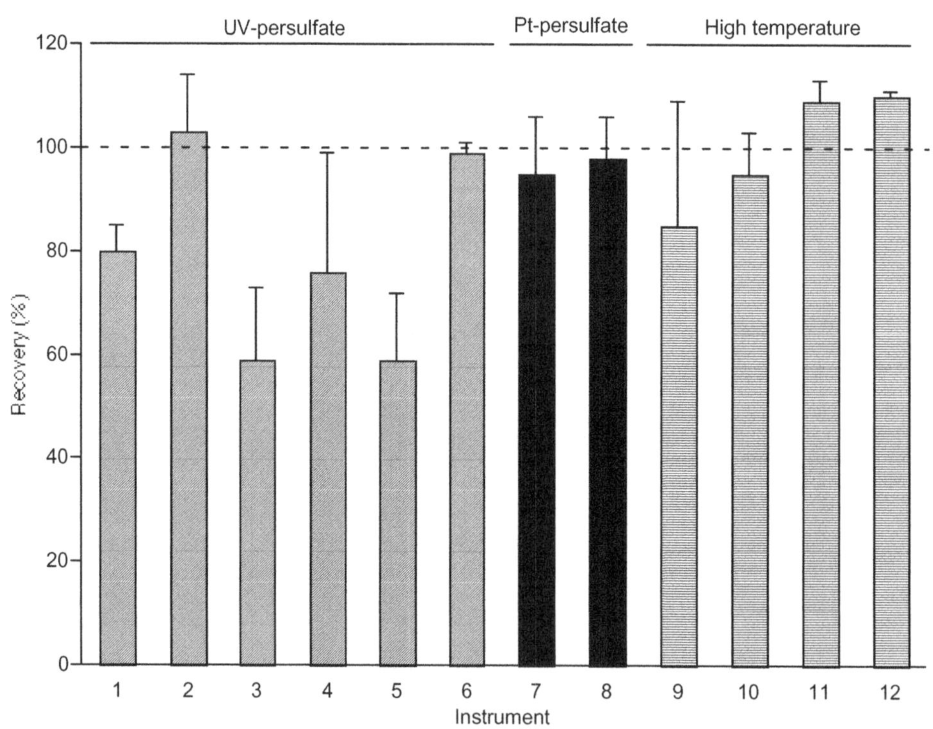

Figure 3.8 Recovery of dodecylbenzene sulfonic acid analyses separated by TOC instrument

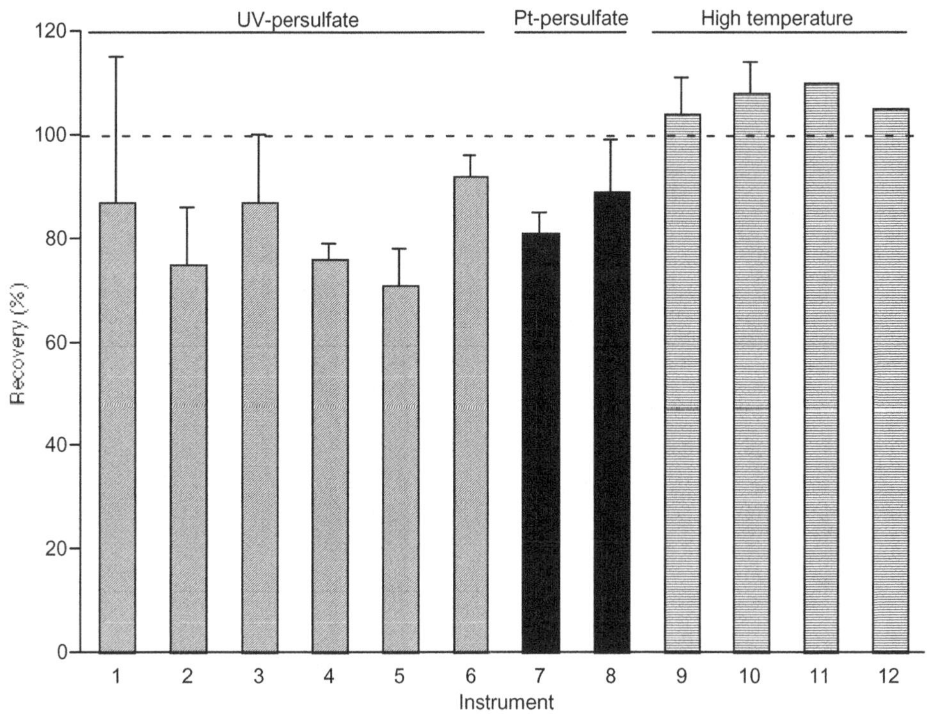

Figure 3.9 Recovery of caffeine analyses separated by TOC instrument

Naturally occurring compounds, such as humic and fulvic acid, were not as problematical for the analysts, with an average recovery of 92 ± 12% for humic acid, and 92 ± 9% for the humic/fulvic acid mixture. Two of the analysts achieved less than 80% recovery with the humic/fulvic acid mixture (range 65 to 107% accuracy), and 4 of the analysts achieved less than 80% recovery with the humic acid alone (range 71 to 117%). The high temperature and the uv-persulfate instruments had the most difficulty with the humic and fulvic acids (Figs. 3.10 and 3.11). The role of the analysts was again prevalent and there were differences in recoveries for different instrument models with each oxidation category. The range of recoveries for both standards with the 9 Shimadzu 5000A instruments was 65 to 117%, 81 to 96% with the 6 Phoenix 8000 instruments, and 71 to 107% with the 6 DC-180 instruments.

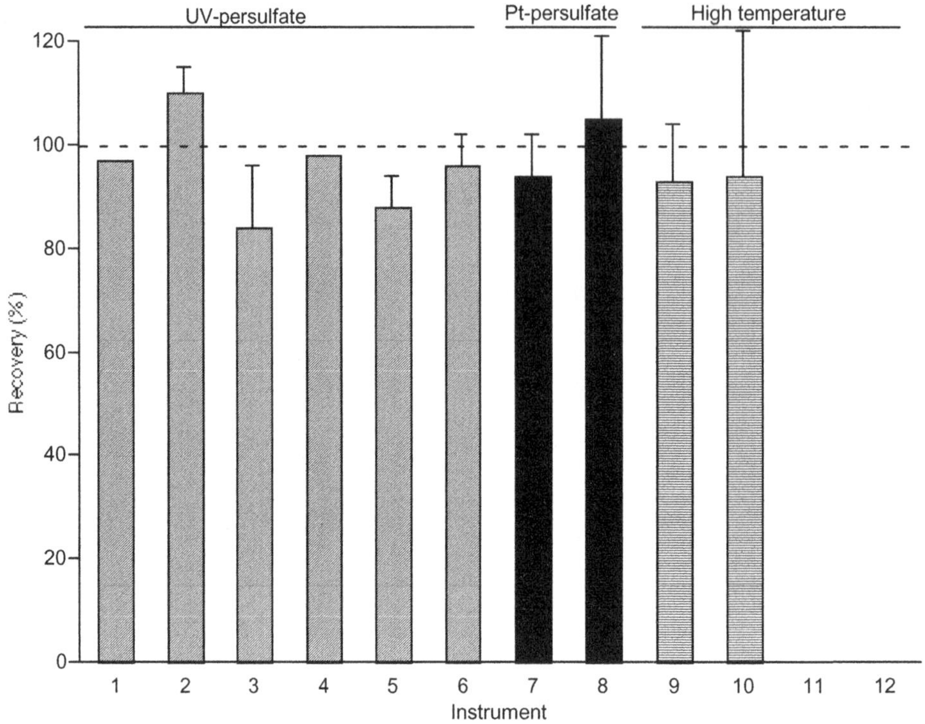

Figure 3.10 Recovery of humic/fulvic acid analyses separated by TOC instrument

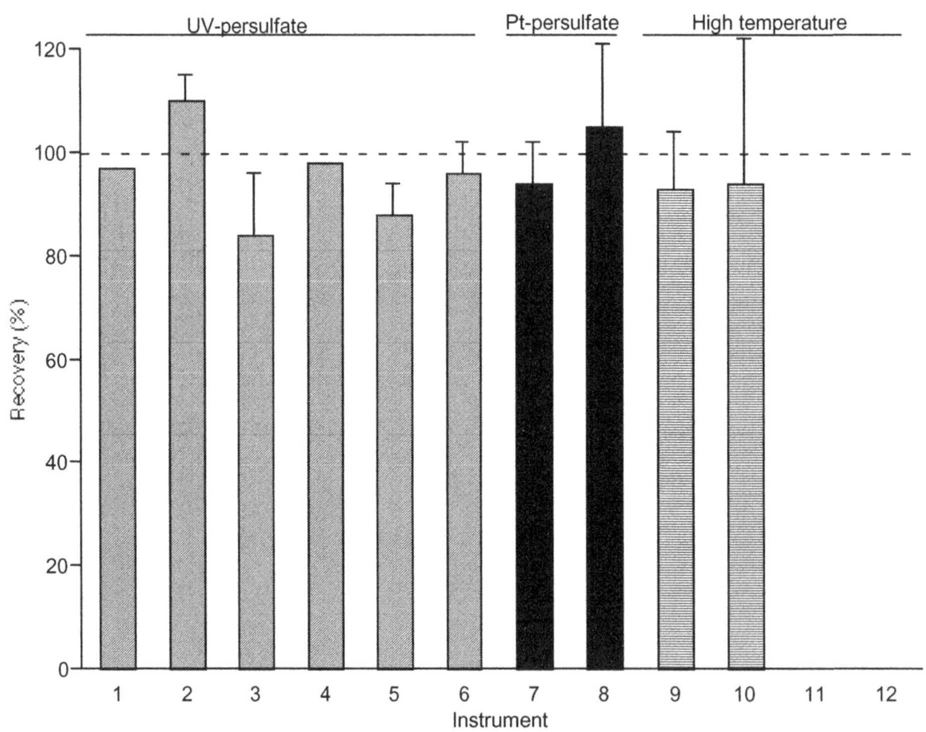

Figure 3.11 Recovery of humic acid analyses separated by TOC instrument

Particles were difficult for nearly all analysts, with the average recoveries and ranges equal to 70 ± 9% (range 54 to 100%) for the leaf powder, 52 ± 31% (range 10 to 133%) for P-17, and 42 ± 34% (range 2 to 116%) for the polystyrene polybeads. The high temperature instruments were superior to the other oxidation technologies with the particles, however there were very broad ranges of performance. The range of recoveries for the high temperature instruments was 58 to 94% with the leaf powder (Fig. 3.12), 15 to 133% with P-17 (Fig. 3.13), and 67 to 118% with the polystyrene polybeads (Fig. 3.14).

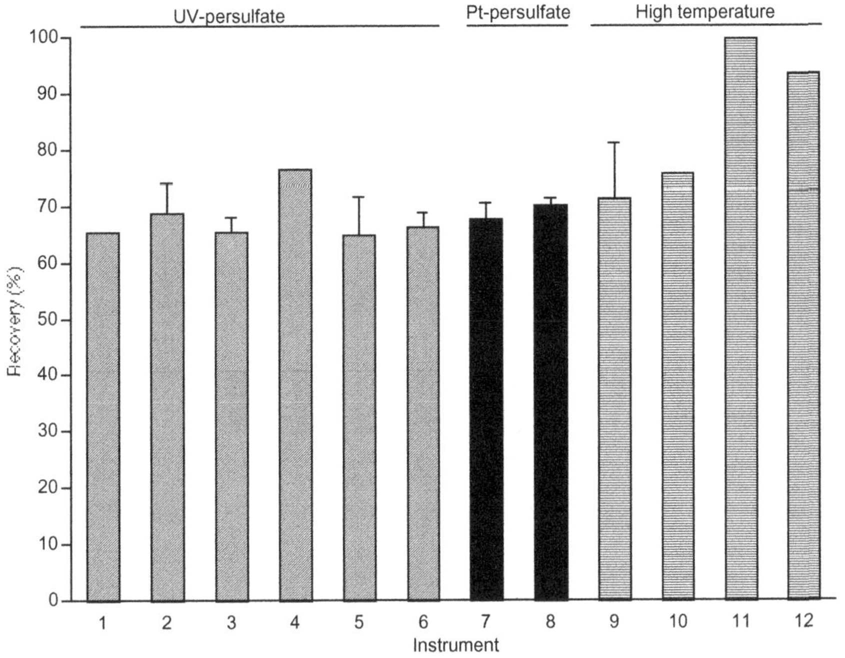

Figure 3.12 Recovery of leaf powder analyses separated by TOC instrument

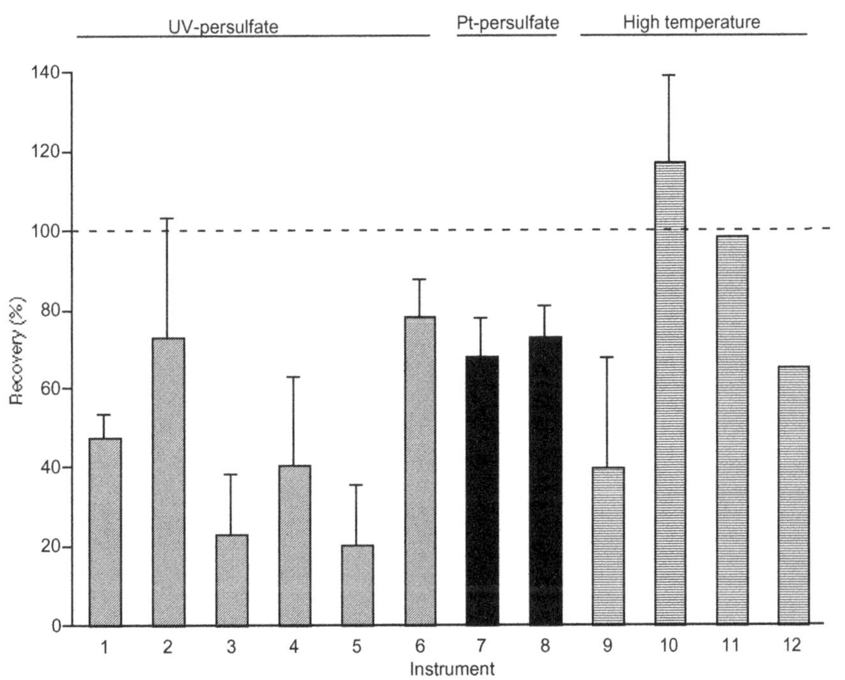

Figure 3.13 Recovery of P-17 analyses separated by TOC instrument

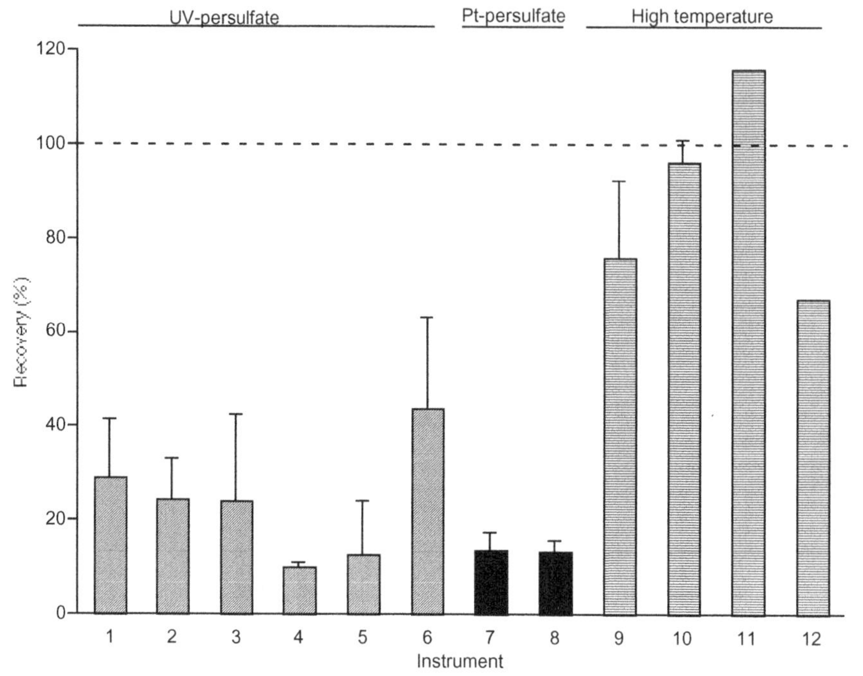

Figure 3.14 Recovery of polybead analyses separated by TOC instrument

Averaged over all instruments and analysts, standards that were solutions were analyzed with 92 ± 9% recovery, while standards that were particles were analyzed with 55 ± 21% recovery. The range for the individual analysts with standards that were solutions was 71 to 109%, and 2 of 39 analysts did not achieve 80% recovery. The range with particles was 22 to 121%, only 3 analysts achieved greater than 80% recovery, and 24 of 39 analysts had less than 50% recovery. When averaged over the individual instruments, all of the instrument models achieved 80% or better recovery with solutions, while only 3 instrument models achieved better than 80% recovery with particles (Fig. 3.15).

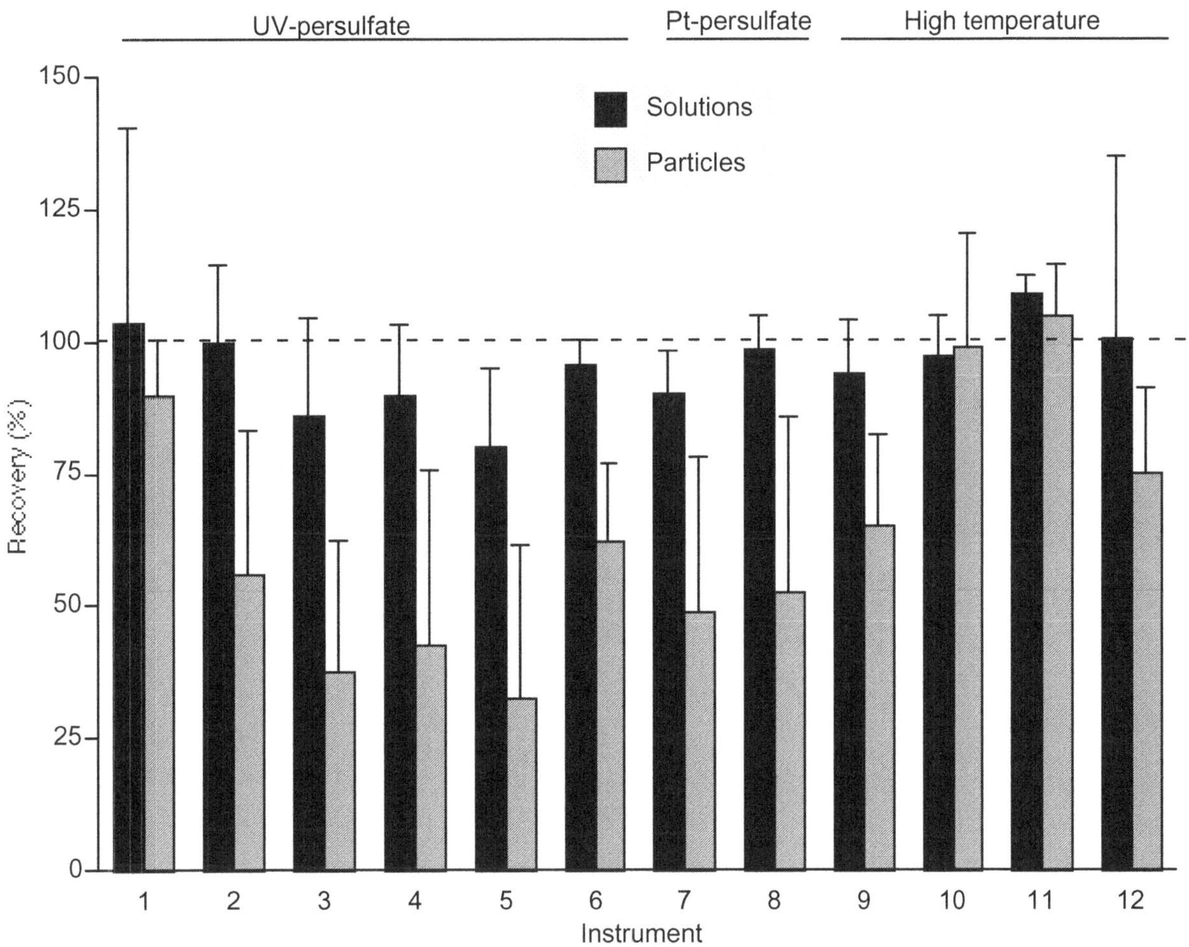

Figure 3.15 Average recovery of analyses of all solutions and particles separated by TOC instrument

Seven of the 12 instrument models were represented by multiple analysts or machines in this study, permitting a fully nested ANOVA. When the statistical test nested analyst within instruments and instruments within methods, we found very few significant differences at either the method or the instrument level. In fact, at the method level, only polystyrene beads provided a significant difference. The average relative analytical errors (RAE) and their standard deviations for the analyses of polystyrene beads were 0.24 ± 0.18 (n = 121) for the high temperature method, 0.72 ± 0.22 (n = 125) for the uv-persulfate method, and 0.87 ± 0.05 (n = 83) for the Pt-persulfate method. Although these differences were significant according to the ANOVA (F = 11.12; P = 0.0037), the Tukey's did not find any significant differences.

At the instrument level, caffeine, dodecylbenzene sulfonic acid, P-17, and some/ of the inorganic carbon amendments provided significant differences (Table 3.10). Of these, only the differences associated with the KHP plus sodium bicarbonate standard were statistically significant with both the ANOVA and the Tukey's.

Most of the error associated with the analysis of the standards was at the level of the individual analyst, and these errors probably prevented us from observing real differences that may exist among instruments and methods. There were 98 different comparisons of standards run by multiple analysts on the same instrument model, and 65 of these revealed significant differences by both the ANOVA and the Tukey's (Table 3.11). These observations support our contention that individual analysts need proficiency evaluation standards that they can use to assess their own performance.

Table 3.10
Instrument-level significant differences

Standard	Method	Ordered Instrument # and RAE* (F; P)†	Tukey's‡ Studentized Range Test
Caffeine	Pt-persulfate	8, 0.098 ± 0.087 (9.23; 0.029)	nsd
		7, 0.214 ± 0.035	(no significant differences)
	Uv-persulfate	6, 0.090 ± 0.021 (4.41; 0.014)	nsd
		5, 0.289 ± 0.065	
		4, 0.240 ± 0.024	
		3, 0.158 ± 0.094	
		2, 0.247 ± 0.112	
		1, 0.263 ± 0.134	
Caffeine + salts + IC	Temp	12, 0.343 ± 0.043 (5.01; 0.031)	nsd
		10, 0.021 ± 0.013	
		11, 0.605 ± 0.119	
		9, 0.099 ± 0.141	
Dodecylbenzene sulfonic acid	Uv-persulfate	6, 0.041 ± 0.027 (9.84; 0.005)	nsd
		5, 0.406 ± 0.142	
		3, 0.411 ± 0.156	
		2, 0.010 ± 0.050	
		1, 0.204 ± 0.040	
Potassium hydrogen phthalate w/$NaHCO_3$	Uv-persulfate	6, 0.071 ± 0.161 (43.63; 0.0001)	2, 1, 4, 6, 5, 3
		5, 0.071 ± 0.036	
		4, 0.084 ± 0.004	
		3, 0.068 ± 0.045	
		2, 0.439 ± 0.075	
		1, 0.280 ± 0.046	
P-17	Uv-persulfate	6, 0.197 ± 0.081 (9.10; 0.007)	nsd
		5, 0.801 ± 0.154	
		4, 0.595 ± 0.181	
		3, 0.771 ± 0.153	
		2, 0.526 ± 0.053	
		1, 0.298 ± 0.255	

* RAE = Relative analytical error; data are ordered as increasing RAE and expressed as $\bar{x}$ ± S.D.
† F= F-value and P = probability from ANOVA
‡ Underscore connects values that are not significantly different

Table 3.11
Machine-level significant differences

Standard	Method	Instrument # (F; P)*	Machine # and statistical† groupings
Caffeine	Uv-persulfate	2 (13.17; 0.0046)	16, 10
		3 (4.80; 0.0052)	9, 11, 17, 29, 28
		5 (9.96; 0.0001)	3, 33, 25, 31, 4, 14
		6 (8.89; 0.0001)	34, 18, 24, 20
	Pt-persulfate	7 (16.25; 0.0001)	19, 22, 13, 21, 5
		8 (0.14; 0.7096)	nsd (no significant differences)
	Temp	9 (14.41; 0.0001)	38, 30, 7, 1, 6, 12, 26, 15, 23
Caffeine + salts + IC	Uv-persulfate	2 (110.47; 0.0005)	10, 16
		3 (10.65; 0.0013)	11, 17, 29, 9, 8
		5 (27.10; 0.0001)	25, 31, 33, 4, 14, 3
		6 (4.09; 0.0471)	nsd
	Pt-persulfate	7 (252.52; 0.0001)	13, 5, 21, 19, 22
		8 (84.63; 0.0008)	8, 35
	Temp	9 (4.84; 0.0004)	38, 7, 26, 6, 1, 12, 23, 30, 15
Dodecylbenzene sulfonic acid	Uv-persulfate	2 (0.86; 0.3678)	nsd
		3 (2.96; 0.0312)	9, 11, 17, 28, 29
		5 (12.83; 0.0001)	3, 25, 33, 14, 4, 31
		6 (0.56; 0.6434)	nsd

(continued)

Table 3.11 (continued)
Machine-level significant differences

Standard	Method	Instrument # (F; P)*	Machine # and statistical† groupings
	Pt-persulfate	7 (25.49; 0.0001)	5, 21, 19, 22, 13
		8 (3.49; 0.0781)	nsd
	Temp	9 (25.65; 0.0001)	15, 6, 1, 26, 30, 23, 7, 12, 38
Fulvic acid + salts	Uv-persulfate	2 (0.96; 0.3822)	nsd
		3 (19.65; 0.0001)	17, 11, 29, 28, 9
		5 (13.69; 0.0002)	3, 14, 31, 33, 4, 25
		6 (2.45; 0.1100)	20, 24, 18, 34
	Pt-persulfate	7 (3.59; 0.0191)	19, 22, 5, 21, 13
		8 (0.21; 0.6691)	nsd
		9 (14.22; 0.0001)	15, 7, 30, 26, 1, 23, 38, 6, 12
Humic/fulvic acid mixture	Uv-persulfate	2 (0.78; 0.4271)	nsd
		3 (2.96; 0.0312)	9, 11, 17, 28, 29
		5 (42.91; 0.0001)	3, 31, 25, 33, 4, 14
		6 (2.88; 0.0984)	34, 24, 20
	Pt-persulfate	7 (24.44; 0.0001)	5, 13, 19, 21, 22
		8 (1.14; 0.3458)	nsd
	Temp.	9 (10.06; 0.0001)	15, 1, 26, 6, 38, 30, 23, 12, 7

(continued)

Table 3.11 (continued)
Machine-level significant differences

Standard	Method	Instrument # (F; P)*	Machine # and statistical† groupings
Humic/fulvic acid mixture + IC	Uv-persulfate	2 (14.95; 0.0180)	10, 16
		3 (1.17; 0.3796)	nsd
		5 (13.72; 0.0001)	31, 4, 33, 3, 14, 25
		6 (3.50; 0.0667)	nsd
	Pt-persulfate	7 (2.43; 0.0738)	nsd
		8 (430.19; 0.0001)	8, 35
	Temp.	9 (4.72; 0.0005)	38, 15, 1, 7, 26, 6, 23, 12, 30
Humic acid	Uv-persulfate	2 (4.94; 0.0903)	nsd
		3 (51.69; 0.0001)	9, 11, 28, 29, 17
		5 (10.04; 0.0002)	3, 25, 4, 31, 33, 14
		6 (1.12; 0.3381)	nsd
		7 (17.38; 0.0001)	21, 5, 22, 13
		8 (4.52; 0.0711)	nsd
		9 (1.46; 0.1933)	nsd
Potassium hydrogen phthalate	Uv-persulfate	2 (0.07; 0.7983)	nsd
		3 (13.75; 0.0001)	9, 11, 17, 29, 28
		5 (12.13; 0.0001)	3, 31, 25, 14, 33, 4
		6 (0.97; 0.4148)	nsd
	Pt-persulfate	7 (1.58; 0.1868)	nsd
		8 (1.39; 0.2554)	nsd
	Temp.	9 (3.51; 0.0010)	15, 1, 30, 6, 26, 7, 23, 12, 38
Potassium hydrogen phthalate + IC	Uv-persulfate	2 (31.72; 0.0049)	10, 16
		3 (6.64; 0.0071)	9, 29, 11, 17, 28
		5 (7.44; 0.0022)	3, 25, 4, 31, 14, 33
		6 (2.27; 0.1493)	nsd

(continued)

Table 3.11 (continued)
Machine-level significant differences

Standard	Method	Instrument # (F; P)*	Machine # and statistical† groupings
	Pt-persulfate	7 (128.63; 0.0001)	13, 5, 21, 19, 22
		8 (901.04; 0.0001)	8, 35
	Temp.	9 (6.94; 0.0001)	38, 7, 23, 1, 12, 26, 6, 15, 30
Potassium hydrogen phthalate + salts	Uv-persulfate	2 (20.80; 0.0103)	10, 16
		3 (0.59; 0.6756)	nsd
		5 (23.68; 0.0001)	3, 31, 14, 25, 33, 4
		6 (0.67; 0.5880)	nsd
	Pt-persulfate	7 (30.13; 0.0001)	13, 21, 5, 22, 19
		8 (0.73; 0.4396)	nsd
	Temp.	9 (0.44; 0.8874)	nsd
Leaf powder	Uv-persulfaate	2 (6.69; 0.0609)	nsd
		3 (3.37; 0.0542)	nsd
		5 (7.34; 0.0039)	3, 25, 4, 31, 33, 14
		6 (6.79; 0.0054)	24, 18, 20, 34
	Pt-persulfate	7 (4.47; 0.0073)	19, 5, 21,22, 13
		8 (5.52; 0.0786)	nsd
	Temp.	9 (9.80; 0.0001)	15, 7, 1, 30, 23, 6, 26, 38, 12
P-17	Uv-persulfate	2 (55.90; 0.0001)	16, 10
		3 (2.39; 0.0778)	nsd
		5 (4.47; 0.0033)	33, 3, 4, 14, 31, 25
		6 (5.87; 0.0025)	20, 18, 24, 34
	Pt-persulfate	7 (2.91; 0.0274)	19, 5, 21, 13, 22
		8 (24.15; 0.0003)	35, 8
	Temp.	9 (80.06; 0.0001)	6, 26, 23, 15, 30, 1, 12, 38, 7

(continued)

Table 3.11 (continued)
Machine-level significant differences

Standard	Method	Instrument # (F; P)*	Machine # and statistical† groupings
Polystyrene beads	Uv-persulfate	2 (20.70; 0.0011)	10, 16
		3 (8.59; 0.0002)	11, 17, 28, 9, 29
		5 (10.45; 0.0001)	3, 33, 4, 31, 25, 14
		6 (2.62; 0.0948)	34,18, 24, 20
	Pt-persulfate	7 (5.87; 0.0004)	19, 22, 5, 21, 13
		8 (0.06; 0.8149)	nsd
	Temp.	9 (1.96; 0.0600)	nsd
Polystyrene beads + salts	Uv-persulfate	2 (5.36; 0.0816)	nsd
		3 (10.23; 0.0015)	28, 17, 11, 29, 9
		5 (26.68; 0.0007)	33, 4, 31, 14
		6 (2.62; 0.0948)	34, 18, 24, 20
		7 (35.42; 0.0001)	13, 5, 22, 21, 19
		8 (4.74; 0.0951)	nsd
		9 (25.27; 0.0001)	15, 7, 30, 38, 23, 12, 26, 6, 1

* F = F-value and P = probability from ANOVA
† Machine listing in order of decreasing Relative Analytical Error and the underscore connect groupings that are not significantly different according to the Tukey's Studentized Range Test.

The data obtained with the proficiency evaluation standards confirmed our concerns about the efficacy of the U.S. EPA certification guidelines for TOC analysis. Certification based upon easily oxidized solutions such as KHP would be readily achieved by most analysts (98% in this study), but it would not be a good indication of TOC analytical performance with other solutions (26% failure rate for caffeine and 38% failure rate for dodecylbenzene sulfonic acid). The inclusion of an IC amendment to the KHP standard also caused 26% of analysts to achieve less than 80% accuracy. Most especially, KHP would not indicate performance with the suspended organic constituents within a water sample (85 to 92% failure rate with particles). If anything, our standards with particles might overestimate analytical accuracy expected with natural water samples because the particles we used were relatively small and uniform. In natural waters the particulate load is dominated by inorganic particles, but all inorganic surfaces in surface waters should have some amount of organic sorption to the surface. Turbid waters such as small rivers during storms, or larger rivers under even low flow conditions, would present a wide range of particle sizes to the TOC analyst, making representative sampling and subsequent delivery of sample to an oxidation zone more difficult than working with our standards.

INFLUENCE OF PARTICLES ON DBPFP

Our data suggest that particles can compete with dissolved precursor material in reactions with applied disinfectants and contribute DBP precursors. The role of particles in DBPFP, however, seems to be related to overall chlorine demand. DBP formation was generally reduced when the DBPFP tests were chlorine limited and increased when the tests were precursor limited (i.e., chlorine demand was substantially met). Additional complexity influencing the role of particles includes the particle sources, season, and concentration, especially during storms that generate additional runoff that is typically high in both DOC and SOC.

Five sets of DBPFP tests were conducted on two raw water sources: Crum Creek and Neshaminy Creek in April and August 1998 under baseflow conditions, and Neshaminy Creek in June 1999 during a storm. Dissolved and suspended organic carbon and total suspended solids concentrations were higher in Crum Creek than Neshaminy Creek under baseflow conditions (Table 3.12). The June storm sample from Neshaminy Creek exhibited elevated DOC, SOC, and suspended solid concentrations (Table 3.12). The SOC:DOC ratio ranged from 0.14 to 0.70 at baseflow, and increased dramatically to 1.58 during the storm. A comparison of the suspended organic carbon concentrations determined by TOC analysis with high temperature combustion and those determined by gravimetric analysis showed that TOC consistently underestimated the SOC concentrations, but to a variable degree (Table 3.12). This underscores the problems previously reported for handling suspended organic carbon loads.

Table 3.12
DOC and SOC data* for samples used in DBPFP analyses.

		TOC analysis		Gravimetric analysis		
Site	Date	DOC	SOC†	SOC‡	Suspended Solids	% SOC
Crum Creek	4/20/98	3.161 ± 0.171 (5)	0.324 ± 0.247 (5)	2.21	16.99	13.0
Neshaminy Creek	4/20/98	3.027 ± 0.139 (5)	0.057 ± 0.152 (5)	1.01	10.24	9.9
Crum Creek	8/11/98	5.331 ± 0.014 (3)	1.327 + 0.021 (3)	2.12	15.33	13.8
Neshaminy Creek	8/10/98	3.740 ± 0.092 (3)	0.013 ± 0.181 (3)	0.51	3.37	15.1
Neshaminy Creek	6/2/99	5.708 ± 0.196 (3)	6.097 ± 0.211 (2)	9.02 ± 1.51 (3)	148.50 ± 27.28 (3)	6.1

* Data reported as $\overline{x}$ ± SD (n) in units of mg/L.
† Determined as the difference TOC-DOC
‡ Assumes that ash free dry mass (AFDM) is 50% C

Chlorine demand curves generated as part of the DBPFP test protocol can be used to separate chlorine-limited from precursor-limited reactions. The slope of the chlorine dose versus total chlorine residual line approaches one as the chlorine demand is met (precursor limitation). The filtered water samples from both sites in April (Fig. 3.16) and from the Neshaminy site in August (Fig. 3.17) exhibited precursor limitation, while the filterered and unfiltered samples from Crum Creek in August (Fig. 3.17) and the Neshaminy samples from June (Fig. 3.18) show chlorine limitation. In these latter samples, the chlorine demand curves indicate that the chlorine demand had not been fully met even at the highest applied chlorine doses.

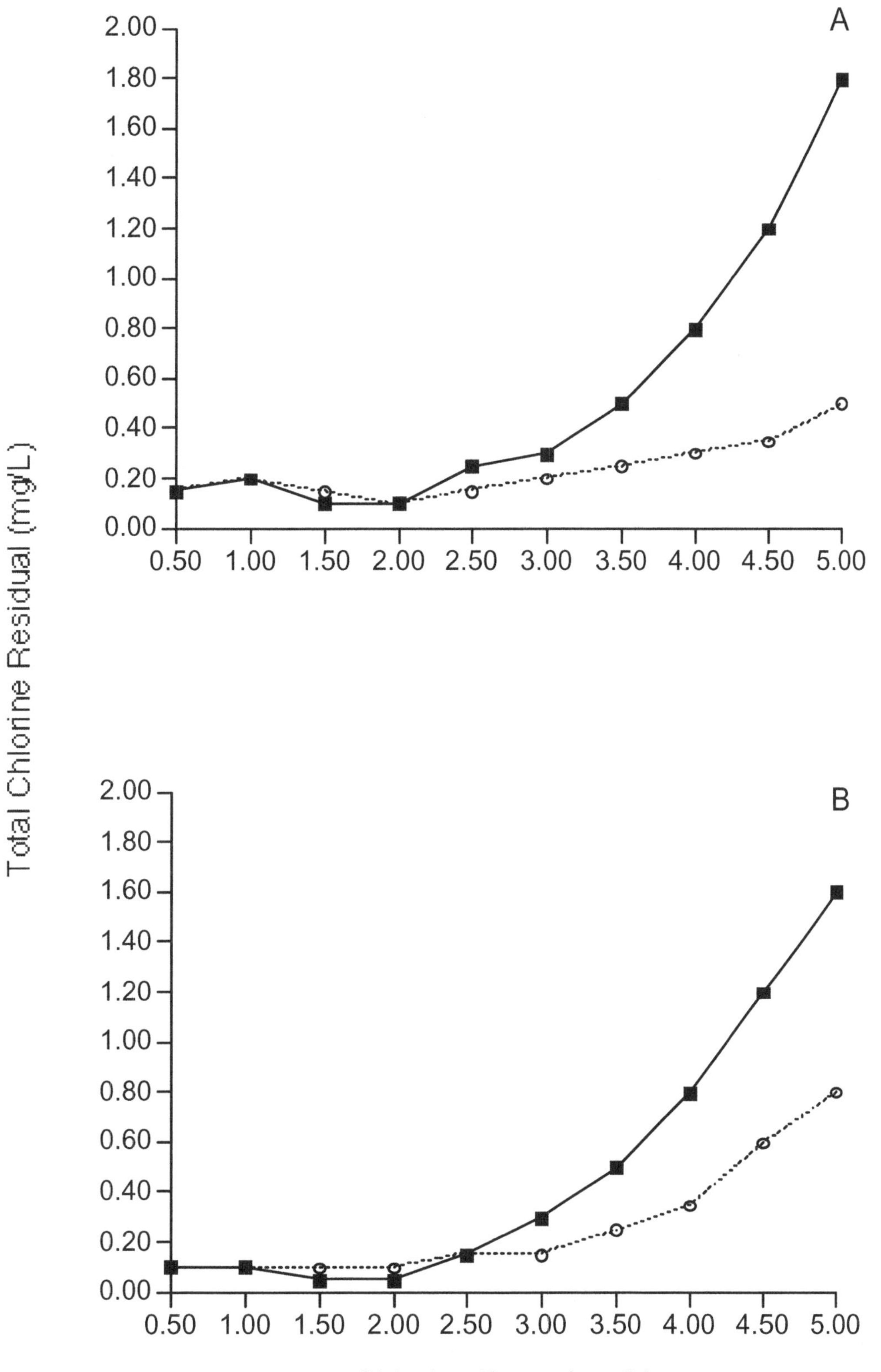

Figure 3.16 Chlorine demand of raw and filtered waters sampled in April 1998. A. Crum Creek. B. Neshaminy Creek. E raw, B filtered.

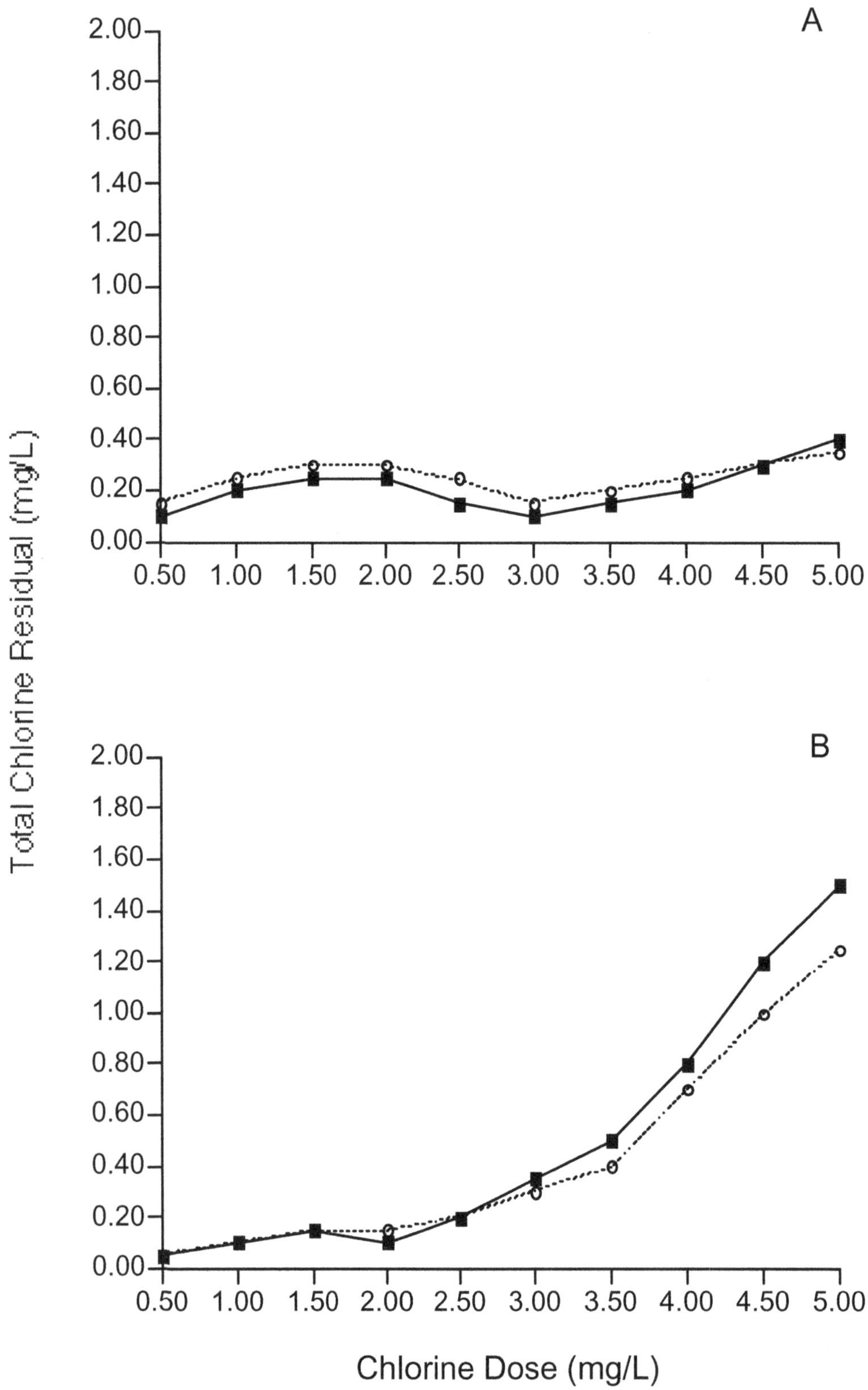

Figure 3.17 Chlorine demand of raw and filtered waters sampled in August 1998. A. Crum Creek. B. Neshaminy Creek. E raw, B filtered.

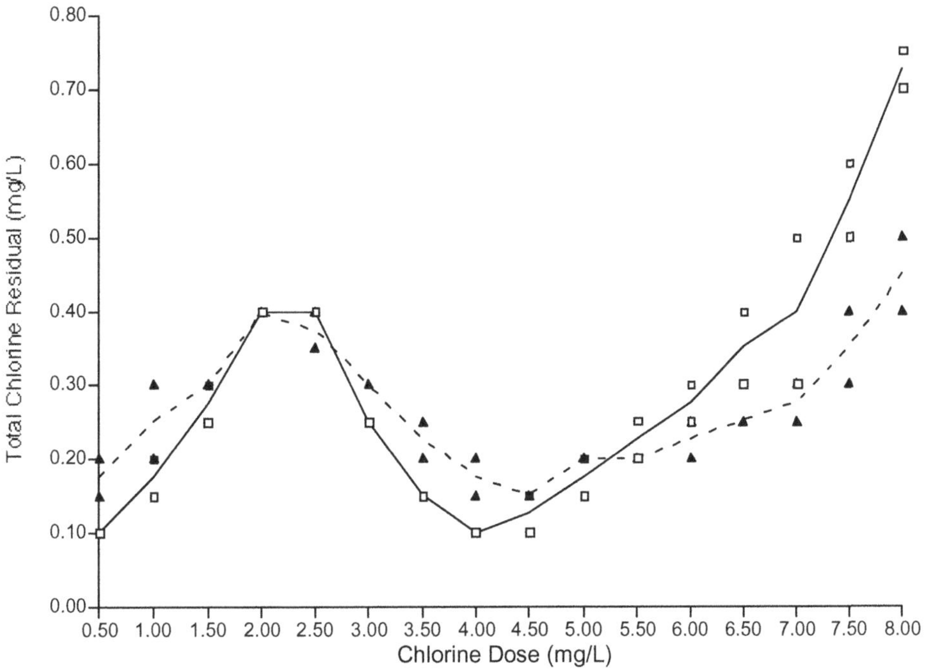

Figure 3.18 Chlorine demand of raw and filtered waters from Neshaminy Creek sampled in June 1999. H raw, -- raw average, G filtered, — filtered average

The initial experiment in April did not use the same chlorine dose for the filtered and unfiltered samples at the lower chlorine doses (Table 3.13) so these data are excluded for the evaluation of particle influences. However, for the higher doses in April and the doses in both August (Table 3.14) and June (Table 3.15), filtered and unfiltered samples were dosed identically. Therefore, we can calculate the influence of particles on DBPFP, as a percentage of the dissolved precursors by the following calculation:

$$[(DBP_{\text{unfiltered}} - DBP_{\text{filtered}})/\ DBP_{\text{unfiltered}}]100.$$

In nearly all samples considered precursor limited, particles increased the DBP formation, while in chlorine-limited samples the opposite was true (Table 3.15). The only exception was the HAA results for the August 1998 Crum Creek samples exposed to the higher chlorine dose. At an applied chlorine dose of 2.5 mg/L, where the reactions would certainly have been chlorine limited, the HAAs were significantly higher in the filtered sample, as expected. However, at 5 mg/L applied chlorine, the reverse was true, and the HAAs in the filtered sample were lower than in the unfiltered sample.

Table 3.13
Contribution of particles to DBPFP, April 20, 1998.

Site	Treatment	Dose (mg/L)	DBP (μg/L) HAA5	HAA6	Chloral hydrate	THM
Crum Creek	Filtered	2.5	29.17	33.03	5.32	37.21
Crum Creek	Unfiltered	3.0	48.92	52.17	7.85	78.42
Crum Creek	Filtered	5.0	90.51	95.55	7.42	105.77
Crum Creek	Unfiltered	5.0	102.72	106.18	26.63	114.51
Neshaminy	Filtered	2.5	48.70	52.10	4.64	73.85
Neshaminy	Unfiltered	3.0	69.87	74.12	8.44	87.68
Neshaminy	Filtered	5.0	91.45	94.63	8.06	91.83
Neshaminy	Unfiltered	5.0	111.67	115.59	11.82	110.03

Table 3.14
Contribution of particles to DBPFP, August 10 - 11, 1998.

Site	Treatment	Dose (mg/L)	DBP (μg/L) HAA5	HAA6	Chloral hydrate	THM
Crum Creek	Filtered	2.5	22.73	25.29	3.62	44.23
Crum Creek	Unfiltered	2.5	18.66	20.35	2.61	42.66
Crum Creek	Filtered	5.0	62.62	66.04	22.47	171.26
Crum Creek	Unfiltered	5.0	75.40	78.96	21.01	138.66
Neshaminy	Filtered	2.5	15.36	17.84	5.15	93.13
Neshaminy	Unfiltered	2.5	25.06	31.36	6.44	81.40
Neshaminy	Filtered	5.0	32.11	36.20	10.11	97.00
Neshaminy	Unfiltered	5.0	56.80	63.65	11.92	113.22

Table 3.15
Contribution of particles to DBPFP, June 2, 1999.

Site	Treatment	Dose (mg/L)	DBP (μg/L) HAA5	HAA6	Chloral hydrate	THM
Neshaminy	Filtered	3.0	31.99 ± 0.43	32.75 ± 0.45	1.51 ± 0.17	43.87 ± 0.85
Neshaminy	Unfiltered	3.0	30.97 ± 1.40	31.85 ± 1.62	1.26 ± 0.06	32.73 ± 1.62
Neshaminy	Filtered	5.0	76.44 ± 0.03	77.92 ± 0.21	7.57 ± 0.08	95.90 ± 2.79
Neshaminy	Unfiltered	5.0	62.20 ± 1.80	65.34 ± 2.03	4.54 ± 0.60	76.94 ± 4.62
Neshaminy	Filtered	6.0	113.36 ± 3.05	115.41 ± 3.25	15.20 ± 0.20	136.20 ± 1.01
Neshaminy	Unfiltered	6.0	104.81 ± 0.82	106.17 ± 1.00	12.18 ± 0.16	114.08 ± 1.28
Neshaminy	Filtered	8.0	182.40 ± 1.24	184.44 ± 1.24	26.70 ± 0.74	184.63 ± 2.84
Neshaminy	Unfiltered	8.0	181.54 ± 3.22	183.21 ± 2.91	24.66 ± 0.42	172.12 ± 5.70

The role of suspended organic material is even more complex when results for individual DBP compounds are examined. For example, more chloral hydrate was generally formed from those samples where the suspended organic material probably was derived from natural sources such as algae, as opposed to anthropogenic sources such as sewage outfalls. This can be seen by comparing the chloral hydrate concentrations produced by the 5 mg/L applied chlorine doses for Crum vs. Neshaminy in April and August, and the Neshaminy results from April and August compared to those from June (Tables 3.13-3.15). Furthermore, in the April 1998 Crum and Neshaminy samples where removal of particles had the most pronounced effect on the chlorine demand curves, filtering substantially reduced the formation of chloral hydrate. The results suggest that chloral hydrate precursor material was preferentially associated with particles in these samples. However, this does not hold for the results from the June 1999 Neshaminy sample. This sample, reflecting a runoff event, produced chloral hydrate levels slightly higher in the filtered samples than in the unfiltered sample.

Because conditions are controlled and independent of water treatment practices, and because the protocol simultaneously defines DBP formation and the chlorine demand curve for a sample, the DBPFP test is a useful tool for investigating DBP formation characteristics of water samples. The DBPFP analyses for the June 1999 Neshaminy sample were performed with replication. Estimates of the precision of the DBPFP assay ranged from less than 1% to approximately 5%, significantly less than most of the differences observed between filtered and unfiltered samples.

The assessment of particle contributions to DBPFP was a small component of this research effort. Particle concentration and organic content were the only characterizations made, and experimental manipulation was limited to the removal of particles prior to chlorination. Our data suggest that the role of particles in DBP formation is complex and potentially significant. More research is needed to ascertain the origin and composition of particles that may affect DBP formation, and the mechanisms of these effects within the context of conventional water treatment practices.

CHAPTER 4
SUMMARY AND RECOMMENDATIONS

A test of TOC analytical accuracy was performed using proficiency evaluation standards prepared at the Stroud Water Research Center. Participants from 33 laboratories that operated a total of 39 TOC analyzers, included 19 water utility laboratories, 5 academic laboratories, 2 government laboratories, 2 private laboratories, and 5 TOC instrument manufacturers. Twelve different TOC analyzer models were represented, including 6 based on uv-persulfate oxidation, 2 based on Pt-persulfate oxidation, and 4 based on high-temperature combustion. Standards were prepared to test detector calibration, the ability to separate inorganic C from organic C, the recovery and analytical accuracy when challenged with solutions of chemically recalcitrant molecules and suspensions of particles, and the influence of matrix modification on these capabilities.

Based on the results from all laboratories, the highest analytical accuracy was achieved with KHP, an easily oxidized solution typical of that used by the U.S. EPA to qualify analysts for TOC measurements. Analytical difficulties were revealed by difficult to oxidize solutions of caffeine and dodecylbenzene sulfonic acid, particles including freeze dried bacteria, leaf powders, and polystyrene polybeads, and additions of inorganic carbon. Taking the criterion of +/- 20% accuracy used by the U.S. EPA to qualify analysts, 98% of all analysts were qualified based on KHP in deionized water, 74% based on KHP amended with IC, 74% based on caffeine, 62% based on dodecylbenzenesulfonic acid, and a maximum of 15% based on particles. Most of the instrument models tested performed similarly with solutions, but the high temperature systems performed better with particles. Overall, operator performance was a more critical parameter than either oxidation method or instrument model. Our research suggests that either caffeine or dodecylbenzene sulfonic acid would be good model compounds for operators to assess the oxidation effiency of their instruments. Caffeine is perhaps a little easier to handle because dodecylbenzenesulfonic acid is a surfactant, and it tends to form bubbles when concentrated solutions are mixed. Additionally, we have demonstrated that the ability to transport particles from a suspension to the oxidation zone of TOC analyzers contributes to the error when measuring SOC. This suggests the need for different instrument designs if SOC is to be included in the TOC measurement. These results indicate that the standard used by the U.S. EPA to qualify analysts overestimates the accuracy of TOC analyses that would be achieved with natural waters, especially surface waters, used as drinking water supplies.

Our research also explored the influence of naturally occurring particles of disinfection by-product formation. The results suggest that particles can compete with dissolved precursor material in reactions with applied disinfectants and contribute DBP precursors. The role of particles in DBPFP, however, seems to be related to overall chlorine demand. The data were equivocal, with clear increases in DBP's associated with one sampling date, but other samples showing a decrease or no change in by-product formation. The greatest influence of particles was observed in a single reservoir sample that had a 3.6-fold increase in chloral hydrate formation and a 2.1-fold increase in THM formation. These data suggest that the quality and quantity of particles are important variables to consider when evaluating DBPFP with raw waters.

From these results, we make the following recommendations:

- there is a need for appropriate proficiency evaluation standards that can allow the analyst to easily assess their own performance with TOC analyses
- routine TOC instrument evaluation should include inorganic carbon removal, oxidation efficiency of difficult to oxidize solutions, and recovery efficiency of particles
- given the difficulty of handling particles in TOC analysis, the industry should consider the importance of organic particle data and assess alternative methodologies for particle evaluation
- additional research is needed to further explore the role of particles, and address at least two critical questions: (1) do particle-generated DBP's remain particle associated during treatment, and (2) what attributes of particles enhance their role as DBP precursors

APPENDIX A:
ANALYST'S QUESTIONNAIRE

AwwaRF Project 452: TOC Analytical Accuracy

Please Answer Any Questions That Are Applicable To You

1.) Instrument and model number
____.
____.
____.

2.) Sample handling:

- ☐ manufacturer's autosampler
- ☐ other autosampler
  (please identify)_____________________________.
- ☐ manual injection

3.) Please answer the following questions about sample size:

a.) Vial size used by instrument______________________.
b.) Volume of sample in vial__________________________.

4.) Sparging:

- ☐ external
- ☐ internal
- ☐ none

5.) Please specify the following:

a.) Sparge gas used_________________________________.
b.) The gas flow rate________________________________.
c.) The sparge time_________________________________.

6.) Please answer the following:

a.) Is the sample acidified? __Y __N
b.) If so, with what acid?____________________________.
c.) To what final pH?_______________________________.
d.) Is the acid addition done by the instrument__. or operator__?

7.) Volume of sample injected
____.
____.

____.

8.) Analysis Mode:

☐ TC-IC

☐ NPOC

☐ other

(please specify)____________________________________.

9.) Shimadzu users, do you use:

☐ Standard needle injection

☐ SP kit injection

10.) Number of syringe washes or rinses between analyses

____.
____.
____.

11.) Catalyst type

____.
____.
____.

12.) Was the sample stirred during analysis? __Y __N

If yes, then:

☐ by the instrument or autosampler

☐ by some manual method

(please describe)__________________________________.

__________________________________.

13.) Any additional comments on sample handling or sample analysis

____.
____.
____.
____.

REFERENCES

Aiken, G.R. 1992. Chloride Interference in the Analysis of Dissolved Organic Carbon by the Wet Oxidation Method. *Environmental Science and Technology* 26:2435-2439.

American Public Health Association. 1995. *Standard Methods for The Examination of Water and Wastewater*, 19th ed. American Public Health Association, New York.

American Public Health Association. 1996. *Standard Methods for The Examination of Water and Wastewater*, 19th ed Supplement. American Public Health Association, New York.

Averett, R.C., J.A. Leenheer, D.M. McKnight, and K.A. Thorn. 1989. Humic Substances in the Suwannee River, Georgia: Interactions, Properties, and Proposed Structures. U. S. Geological Survey Open-File Report 87-557.

Bellamy, W.D., and M.J. McGuire. 1994. Report From Expert Workshop on Microbial and Disinfection By-Product Research Needs, Miami, Florida, November 3-5, 1993. American Water Works Association Research Foundation.

Benner, R., and J.I. Hedges. 1993. A Test of the Accuracy of Freshwater DOC Measurements by High-Temperature Catalytic Oxidation and UV-Promoted Persulfate Oxidation. *Marine Chemistry* 41:161-165.

British Standards Institution. 1997. Water Analysis – Guidelines for the Determination of Total Organic Carbon (TOC) and Dissolved Organic Carbon (DOC). European Standard EN 1484.

Golladay, S.W. 1997. Suspended Particulate Organic Matter Concentration and Export in Streams. *Journal North American Benthological Society* 16:122-131.

Hobbie, J. E., R. J. Daley and J. Jasper. 1977. Use of Nuclepore Filters for Counting Bacteria by Fluorescence Microscopy. Applied and Environmental Microbiology 33:1225-1228.

Huffman, Jr., E.W.D. amd H.A. Stuber. 1985. Analytical Methodology for Elemental Analysis of Humic Substances. pp. 433-455 In: Aiken, G.R., D.M. McKnight, R.L. Wershaw, and P. MacCarthy (eds.) *Humic Substances in Soil, Sediment, and Water*, John Wiley and Sons.

Kaplan, L. A. 1992. Comparison of High-Temperature and Persulfate Oxidation Methods for Determination of Dissolved Organic Carbon in Freshwaters. *Limnology and Oceanography* 37:1119-1125.

Kaplan, L.A. 1994. A Field and Laboratory Procedure to Collect, Process, and Preserve Freshwater Samples for Dissolved Organic Carbon Analysis. *Limnology and Oceanography* 39:1470-1476.

Kaplan, L. A. 2000. Comparison of Three TOC Methodologies. Journal American Water Works Association 92(4):149-156.

Koprivnjak, J.F., J.G. Blanchette, R.A. Bourbonniere, T.A. Clair, A. Heyes, K.R. Lum, R. McCrea, and T.R. Moore. 1995. The Underestimation of Concentrations of Dissolved Organic Carbon in Freshwaters. *Water Research* 29:91-94.

Malcolm, R. L., and W. H. Durum. 1976. Organic Carbon and Nitrogen Concentrations and Annual Organic Carbon Load of Six Selected Rivers of the United States. U. S. Geological Survey Water Supply Paper 1817-G.

Malcolm, R.L., and P. MacCarthy 1986. Limitations in the use of Commercial Humic Acids in Water and Soil Reserach. *Environmental Science and Technology* 20:904-911.

National Research Council. 1986. *Acid Deposition Long-Term Trends*. National Academy Press, Washington, D.C.

Najm, I.N., J. Marcinko, and J. Oppenheimer. 2000. Evaluating TOC Analytical Results. *JAWWA* 92(8):84-92.

Sharp, J.H. and E.T. Peltzer. 1993. Procedures Subgroup Report. Marine Chemistry 41:37-49.

Sigleo, A.C. 1996. Biochemical Components in Suspended Particles and Colloids: Carbohydrates in the Potomac and Patuxent Estuaries. *Organic Geochemistry* 24:83-93.

Thurman, E.M. 1986. *Organic Geochemistry of Natural Waters*. Martinus Nijhoff/Dr. W. Junk Publishers.

United States Environmental Protection Agency. 1994. Monitoring Requirements for Public Drinking Water Supplies; Proposed Rule. Federal Register 40 CFR Part 141:6331-6444.

Wangersky, P.J. 1990. Sampling for Particulate and Dissolved Material, p. 10-26. In R.S. Wotton [ed.], *The Biology of Particles in Aquatic Systems*. CRC Press.

Water analysis — Guidelines for the Determination of Total Organic Carbon (TOC) and Dissolved Organic Carbon (DOC). British Standard BS EN 1484: 1997.

Windholz, M., S. Budavari, R.F. Blumetti, and E.S. Otterbein. 1983. *The Merck Index,* 10th Edition. Merck & Co., Inc.

ABBREVIATIONS

AFDM	ash free dry mass
ANOVA	analysis of variance
BOM	biodegradable organic matter
C	carbon
°C	degrees Celsius
D/DBP	disinfectants/disinfection by-products
DAPI	4',6-diamidino-2-phenylindole
DBP	disinfectant by-product
DBPFP	disinfectant by-product formation potential
DOC	dissolved organic matter
EDTA	ethylenediamine-tetraacetate (disodium salt)
g	gram
h	hours
HAA5	haloaceticacids
HANs	haloacetonitriles
HPC	heterotrophic plate count
IC	inorganic carbon
IC/L	inorganic carbon per liter
ICR	information collection rule
Kg	kilogram
KHP	potassium hydrogen phthalate
L	liter
L/s	liter per second
mg C/L	milligram carbon per liter
mL	milliliter
mm	millimeter
mM	millimolar
_g/L	microgram per liter
_L/min	microliters per minute
_m	micrometer
_mhos	micromhos
n	number of samples

n.s	not significant
nsd	no significant differences
nm	nanomolar
NOM	natural organic matter
OC	organic carbon
ortho-P	orthophosphate
p	probability of rejecting a tested hypothesis when it is true
P-17	*Pseudomonas fluorescens* strain
PI	propidium iodide
PSWC	Philadelphia Suburban Water Company
Pt	platinum
r	coefficient of linear correlation
r^2	coefficient of determination
RAE	relative analytical error
sd	standard deviation
SOC	suspended organic carbon
SOM	suspended organic matter
TC	total carbon
TOC	total organic carbon
TTHM	trihalomethanes
TYE	tryptone yeast extract
U. S. EPA	United States Environmental Protection Agency
U.S	United States
USGS	United States Geological Survey
Uv	ultraviolet
w/v	weight/volume
x	mean
yr	year

IWA Publishing's authorised EU representative for General Product Safety Regulations is Diane D'Arras, 15 rue Duret, 75116 Paris, France, e-mail: safety@iwap.co.uk.

Printed and bound by CPI Group (UK) Ltd, Croydon, CR0 4YY
17/09/2026
02218799-0009